Statistical Mechanics

Statistical Mechanics

Mateo Mitchell

STATES
ACADEMIC PRESS
www.statesacademicpress.com

Statistical Mechanics
Mateo Mitchell
ISBN: 978-1-63989-496-3 (Hardback)

STATES
ACADEMIC PRESS

Published by States Academic Press,
109 South 5th Street,
Brooklyn, NY 11249, USA

Cataloging-in-Publication Data

Statistical mechanics / Mateo Mitchell.
p. cm.
Includes bibliographical references and index.
ISBN 978-1-63989-496-3
1. Statistical mechanics. 2. Statistical physics. 3. Physics. I. Mitchell, Mateo.
QC174.8 .P75 2022
530.13--dc23

Table of Contents

Preface

Statistical mechanics is a sub-branch of physics which makes use of the principles of statistical methods and probability theory to determine the behavior of macroscopic systems. It attempts to understand the properties of these systems such as heat capacity, pressure, temperature, etc., by studying the microscopic entities which constitute them. Various calculation methods concerned with this subject are Monte Carlo method, perturbation theory and exact solution method. The applications of statistical mechanics include regression analysis of gravitational orbits, weather forecasting and neural networks. This book presents the complex subject of statistical mechanics in the most comprehensible and easy to understand language. It attempts to understand the multiple branches that fall under the discipline of statistical mechanics and how such concepts have practical applications. This book will serve as a valuable source of information for those interested in this field.

To facilitate a deeper understanding of the contents of this book a short introduction of every chapter is written below:

Chapter 1- The mathematical framework which focuses on the application of statistical methods and probability theory to large groups of microscopic entities is termed as statistical mechanics. This chapter has been carefully written to provide an easy introduction to the varied facets of statistical mechanics such as Ising model, Heisenberg model and percolation theory.

Chapter 2- Statistical thermodynamics refers to a theory that predicts the behavior of macroscopic quantities of compounds by analyzing their molecular properties. Some of its important aspects are thermodynamic equilibrium, thermodynamic potentials and statistical foundation of thermodynamics. The topics elaborated in this chapter will help in gaining a better perspective about these facets of statistical thermodynamics.

Chapter 3- Classical ensemble theory is used to provide a probability distribution for the state of the system. A few of its major areas of study within this theory are phase space and its properties, micro-canonical ensemble, canonical ensemble, grand canonical ensemble and Liouville's theorem. This chapter discusses in detail these aspects of classical ensemble theory.

Chapter 4- The state of matter which comprises an ensemble of various non-interacting fermions is termed as a Fermi gas. A few examples of such particles are electrons, protons and neutrons. These particles have a half-integer spin. This chapter has been written to provide a thorough understanding of the different facets of Fermi gas.

Chapter 5- The classical system of interacting particles can be explained by cluster expansion and virial coefficients. Cluster expansion gives a method to understand the logarithm of partition function. In order to completely understand the classical system of interacting particles, it is necessary to understand the processes related to it. The following chapter elucidates the varied processes and mechanisms associated with this area of study.

Finally, I would like to thank the entire team involved in the inception of this book for their valuable time and contribution. This book would not have been possible without their efforts. I would also like to thank my friends and family for their constant support.

Mateo Mitchell

1

Statistical Mechanics: An Introduction

The mathematical framework which focuses on the application of statistical methods and probability theory to large groups of microscopic entities is termed as statistical mechanics. This chapter has been carefully written to pro ide an easy introduction to the aried facets of statistical mechanics such as sing model, eisenberg model and percolation theory.

Statistical mechanics is the branch of physics that combines the principles and procedures of statistics with the laws of both classical and quantum mechanics, particularly with respect to the field of thermodynamics. It aims to predict and explain the measurable properties of macroscopic systems on the basis of the properties and behaviour of the microscopic constituents of those systems. Statistical mechanics interprets, for example, thermal energy as the energy of atomic particles in disordered states and temperature as a quantitative measure of how energy is shared among such particles. Statistical mechanics draws heavily on the laws of probability so that it does not concentrate on the behaviour of every individual particle in a macroscopic substance but on the average behaviour of a large number of particles of the same kind.

Statistical Mechanics as one of the core knowledge domains of physics is a result of persistent efforts made by researchers to understand the microscopic origin of phenomenological theory of macroscopic natural phenomenon called thermodynamics, which has very well established laws based on experimental facts and successful technological applications. These laws are:

- Zeoth law of Themodynamics,
- First Law of Thermodynamics,
- Second Law of Thermodynamics,
- Third law of Thermodynamics.

These laws have been known to us well before:

- The revelation of the fact that matter is made up of atoms or molecules.
- The development of quantum mechanics.

As it became clear that macroscopic bodies are made up of such microscopic particles (atoms, molecules, electrons etc.), it was natural to ask the question, can we explain the behavior of these bodies, which are typically made up of particles of the order of Avogadro's number i.e. $\sim 10^{23}$, starting from the laws governing motion of the microscopic constituents.

Complexity of the Task

The moment one recognizes that a typical macroscopic body say 18 gm of water, is made up of 6.023×10^{23} molecules of water, the problem indeed becomes very complex. Desiring to keep track of all these molecules (their positions and velocities) with time by setting up as many equations of motion and solving these is an impossible task, by the fastest of the super computers available today in a typical life time of a person (it is worth estimating this amount of time required for such an effort by a present day computer!). Lack of means to gather such huge information leaves us with an information crunch also. It is, therefore, natural to believe that it is a hopeless task which inherently has to begin with perhaps too meagre information to arrive at worthwhile conclusions for an enormously complex system.

Statistical Approach

Statistical approach offers a methodology to look at such a complex many body problem to make sense of its dynamical behaviour, bypassing the need to solve equation of motion (Newton's Equation of motion for a classical particle or Schrodinger's Equation of motion for a quantum particle) for each of the constituting particles.

As is characteristic of the statistical approach, it may not provide us knowledge of the behavior of each individual particle, yet it is capable of providing average behavior of important physical quantities of interest in a physical system with certain rules of transaction on energy, number of particles and external conditions etc.

Therefore, statistical mechanics is a way to bring in new concepts besides fundamental laws of classical or quantum mechanics as the case may be. These new concepts involve probability theory via principle of equal apriori probability and are powerful enough to deal with the complexity of the many body problems encountered in nature despite lack of information. The basic question which statistical mechanics tries to address is that can observed properties of a system be obtained from the average behavior of its constituents when these are brought together. Answer indeed leads us to an understanding of macroscopic behavior of a system from microscopic behavior of constituent particles.

Entropy: A Bridge between Thermodynamics and Statistical Mechanics

Entropy of a macroscopic system, which was defined by Rudolf Clausius in a paper published in the Annalen der Physik und Chemie in 1850, is an important measurable physical quantity which offers a bridge to link statistical mechanics (theory) with thermodynamics (experiment).

The meaning of entropy has been a matter of much discussion in thermodynamics in particular and physics in general. The best intuitive interpretation of entropy came from famous Physicist Freeman Dyson while answering the question, what is heat? He described heat of a macroscopic system as nothing but its disordered energy. He said that Its full knowledge requires two numbers, one is a measure of its quantity in say joules and the second a measure of its disorderedness called entropy.

The quantification of entropy happened in thermodynamics through the third law of thermodynamics,

$$dS = \frac{\delta Q}{T}$$

and in statistical mechanics through famous Boltzmann formula:

$$S = k_B \ln \omega$$

Where, S is entropy, Q is heat, T is temperature, k_B is Boltzmann constant and ω is thermo dynamical probability. In nut shell statistical mechanics deals with finding the thermodynamic probability by statistical approach providing a link with other thermodynamic variables via free energy. In thermodynamics free energy, F, is a quantity which can be known by measurements alone, whereas statistical mechanics provides a mean to calculate free energy via Boltzmann formula linking theory with experiments.

Statistical Models

Ising Model

One of the simplest and most famous models of an interacting system is the Ising model. The Ising model was first proposed in Ising's Ph.D thesis and appears in his 1925 paper based on his thesis. In his paper Ising gives credit to his advisor Wilhelm Lenz for inventing the model, but everyone calls it the Ising model. The model was originally proposed as a model for ferromagnetism. Ising was very disappointed that the model did not exhibit ferromagnetism in one dimension, and he gave arguments as to why the model would not exhibit ferromagnetism in two and three dimensions. We now know that the model does have a ferromagnetic transition in two and higher dimensions. A major breakthrough came in 1941 when Kramers and Wannier gave a matrix formulation of the problem. In 1944 Lars Onsager gave a complete solution of the problem in zero magnetic field. This was the first nontrivial demonstration of the existence of a phase transition from the partition function alone.

Consider a lattice of N sites with a spin S on each site. Each spin can take one of two possible values: +1 for spin up and −1 for spin down. There are a total of 2^N possible configurations of the system. A configuration is specified by the orientations of the spins on all N sites: $\{S_i\}$. S_i is the spin on the ith lattice site. The interaction energy is defined to be:

$$E_I\{S_i\} = -\sum_{\langle i,j\rangle} J_{ij} S_i S_j - \sum_{i=1}^{N} B_i S_i$$

where the subscript I represents the Ising model. A factor of 2 has been absorbed into J_{ij} and we set $g\mu_B = 1$ in the last term. $\langle i,j\rangle$ means nearest neighbor pairs of spins. So $\langle i,j\rangle$ is the same as $\langle j,i\rangle$. J_{ij} is the exchange constant; it sets the energy scale. For simplicity, one sets J_{ij} equal to a constant J. If $J > 0$, then the spins want to be aligned parallel to one another, and we say that the interaction is ferromagnetic. If $J < 0$, then the spins want to be antiparallel to one another, and we say that the interaction is antiferromagnetic. If J_{ij} is a random number and can either be positive or negative, then we have what is called a spin glass. For simplicity we will set $J_{ij} = J > 0$ and study the ferromagnetic Ising model. The last term represents the coupling of the spins to an external magnetic field B. The spins are assumed to lie along the z-axis as is the magnetic field $\vec{B} = B\hat{z}$. The spins lower their energy by aligning parallel to the field. I put Bi to indicate the possibility that the field could vary from spin to spin. If the field B_i is random, this is called the random field Ising model. We will assume a constant uniform magnetic field so that $B_i = B > 0$. So the interaction energy becomes,

$$E_I\{S_i\} = -J\sum_{\langle i,j\rangle} S_i S_j - B\sum_{i=1}^{N} S_i$$

The partition function is given by,

$$Z = \sum_{S_1=-1}^{+1}\sum_{S_2=-1}^{+1}\cdots\sum_{S_N=-1}^{+1} e^{-\beta E_I\{S_i\}}$$

One Dimensional Ising Model and Transfer Matrices

Let us consider the one-dimensional Ising model where N spins are on a chain. We will impose periodic boundary conditions so the spins are on a ring. Each spin only interacts with its neighbors on either side and with the external magnetic field B. Then we can write,

$$E_I\{S_i\} = -J\sum_{i=1}^{N} S_i S_{i+1} - B\sum_{i=1}^{N} S_i$$

The periodic boundary condition means that,

$$S_{N+1} = S_1$$

The partition function is,

$$Z = \sum_{S_1=-1}^{+1}\sum_{S_2=-1}^{+1}\cdots\sum_{S_N=-1}^{+1} \exp\left[\beta\sum_{i=1}^{N}(JS_iS_{i+1} + BS_i)\right]$$

Kramers and Wannier showed that the partition function can be expressed in terms of matrices:

$$Z = \sum_{S_1=-1}^{+1}\sum_{S_2=-1}^{+1}\cdots\sum_{S_N=-1}^{+1} \exp\left[\beta\sum_{i=1}^{N}\left(JS_iS_{i+1} + \frac{1}{2}B(S_i + S_{i+1})\right)\right]$$

This is a product of 2 × 2 matrices. To see this, let the matrix P be defined such that its matrix elements are given by,

$$\langle S|P|S'\rangle = \exp\left\{\beta\left[JSS' + \frac{1}{2}B(S+S')\right]\right\}$$

where S and S' may independently take on the values ±1. Here is a list of all the matrix elements:

$$\langle +1|P|+1\rangle = \exp[\beta(J+B)]$$
$$\langle -1|P|-1\rangle = \exp[\beta(J-B)]$$
$$\langle +1|P|-1\rangle = \langle +1|P|-1\rangle = \exp[-\beta J]$$

Thus an explicit representation for P is,

$$P=\begin{pmatrix} e^{\beta(J+B)} & e^{-\beta J} \\ e^{-\beta J} & e^{\beta(J-B)} \end{pmatrix}$$

With these definitions, we can write the partition function in the form:

$$\begin{aligned} Z &= \sum_{S_1=-1}^{+1}\sum_{S_2=-1}^{+1}\cdots\sum_{S_N=-1}^{+1}\langle S_1|P|S_2\rangle\langle S_2|P|S_3\rangle\ldots\langle S_N|P|S_1\rangle \\ &= \sum_{S_1=-1}^{+1}\langle S_1|P^N|S_1\rangle \\ &= \mathrm{Tr}P^N \\ &= \lambda_+^N+\lambda_-^N \end{aligned}$$

where λ_+ and λ_- are the two eigenvalues of P with $\lambda_+ \geq \lambda_-$. The fact that Z is the trace of the Nth power of a matrix is a consequence of the periodic boundary condition Equation $S_{N+1}=S_1$. The eigenvalue equation is,

$$\det\begin{vmatrix} e^{\beta(J+B)}-\lambda & e^{-\beta J} \\ e^{-\beta J} & e^{\beta(J-B)}-\lambda \end{vmatrix} = \lambda^2-2\lambda e^{\beta J}\cosh(\beta B)+2\sinh(2\beta J)=0$$

Solving this quadratic equation for λ gives,

$$\lambda_\pm = e^{\beta J}\left[\cosh(\beta B)\pm\sqrt{\cosh^2(\beta B)-2e^{-2\beta J}\sinh(2\beta J)}\right]$$

When B = 0,

$$\begin{aligned} \lambda_+ &= 2\cosh(\beta J) \\ \lambda_- &= 2\cosh(\beta J) \end{aligned}$$

Now back to the general case with $B \neq 0$. Notice that $\lambda_-/\lambda_+ \leq 1$ where equality is in the case of J = B = 0. In the thermodynamic limit $(N \to \infty)$, only the larger eigenvalue λ_+ is relevant. To see this, we use $(\lambda_-/\lambda_+)<1$ and write the Helmholtz free energy per spin:

$$\begin{aligned} -\frac{F}{Nk_BT} &= \lim_{N\to\infty}\frac{1}{N}\ln Z \\ &= \lim_{N\to\infty}\frac{1}{N}\ln\left\{\lambda_+^N\left[1+\left(\frac{\lambda_-}{\lambda_+}\right)^N\right]\right\} \\ &= \ln\lambda_+ + \lim_{N\to\infty}\frac{1}{N}\ln\left[1+\left(\frac{\lambda_-}{\lambda_+}\right)^N\right] \\ &= \ln\lambda_+ \end{aligned}$$

So the Helmholtz free energy per spin is,

$$\frac{F}{N} = -\frac{k_B T}{N} \ln Z = -k_B T \ln \lambda_+$$
$$= -J - k_B T \ln\left[\cosh(\beta B) + \sqrt{\cosh^2(\beta B) - 2e^{-2\beta J}\sinh(2\beta J)}\right]$$

The magnetization per spin is,

$$m = \frac{M}{N}$$
$$= \frac{1}{\beta N}\frac{\partial \ln Z}{\partial B}$$
$$= -\frac{1}{N}\frac{\partial F}{\partial B}$$

$$= \frac{\sinh(\beta B)}{\sqrt{\cosh^2(\beta B) - 2e^{-2\beta J}\sinh(2\beta J)}}$$

At zero field (B = 0), the magnetization is zero for all temperatures. This means that there is no spontaneous magnetization and the one-dimensional Ising model never exhibits ferromagnetism. The reason is that at any temperature the average configuration is determined by two opposite and competing tendencies: The tendency towards a complete alignment of spins to minimize the energy, and the tendency towards randomization to maximize the entropy. The over-all tendency is to minimize the free energy F = E –TS. For the one-dimensional model the tendency for alignment always loses out, because there are not enough nearest neigbors. However, in higher dimensions, there are enough nearest neighbors and a ferromagnetic transition can occur.

The method of transfer matrices can be generalized to two and higher dimensions, though the matrices become much larger. For example, in two dimensions on an m × m square lattice, the matrices are $2^m \times 2^m$. In 1944, Onsager solved the two dimensional Ising model exactly for the zero field case, and found a finite temperature ferromagnetic phase transition. This is famous and is known as the Onsager solution of the 2D Ising model. No one has found an exact solution for the three dimensional Ising model.

Applications of the Ising Model

The Ising model can be mapped into a number of other models. Two of the better known applications are the lattice gas and the binary alloy.

Lattice Gas

The term lattice gas was first coined by Yang and Lee in 1952, though the interpretation of the model as a gas was known earlier. A lattice gas is defined as follows. Consider a lattice of V sites (V = volume) and a collection of N particles, where N < V. The particles are placed on the vertices of the lattice such that not more than one particle can occupy a given site, and only particles on nearest-neighbor lattice sites interact. The interaciton potential between two lattice sites i and j is given by $V\left(\left|\vec{r}_i - \vec{r}_j\right|\right)$ with:

$$V(r) - \begin{cases} \infty & (r = 0) \\ -\varepsilon_o & (r = a) \\ 0 & \text{otherwise} \end{cases}$$

where a is the lattice spacing. The occupation n_i of a lattice site i is given by,

$$n_i = \begin{cases} 1 & \text{if site i is occupied} \\ 0 & \text{if site i is unoccupied} \end{cases}$$

The interaction energy is,

$$E_G\{n\} = -\varepsilon_o \sum_{\langle i,j \rangle} n_i, n_j$$

We can map this into the Ising model by letting spin-up denote an occupied site and letting spin-down denote an unoccupied site. Mathematically, we write:

$$S_i = 2n_i - 1$$

So $S_i = 1$ means site i is occupied and $S_i = -1$ means site i is unoccupied. One can then map the lattice gas model into Ising model. For example, by comparing the partition functions, it turns out that $\varepsilon_o = 4J$.

Binary Alloy

A binary alloy is a solid consisting of 2 different types of atoms. For example, β–brass is a body-centered cubic lattice made up of Zn and Cu atoms. At T = 0, the lattice is completely ordered and a copper atom is surrounded by zinc atoms and vice-versa. However, at non-zero temperatures the zinc and copper atoms can exchange places. Above a critical temperature of T = 742 K, the Zn and Cu atoms are thoroughly mixed so that the probability of finding a Zn atom on given site is 1/2. Similarly the probability of finding a Cu atom on given site is 1/2. To model a binary alloy, one starts with a lattice of N sites, and two different types of atoms, A and B. Each site has only one atom so that $N_A + N_B = N$. The occupation of each site is,

$$n_i = \begin{cases} 1 & \text{if site i is occupied by atom A} \\ 0 & \text{if site i is occupied by atom B} \end{cases}$$

There are interaction energies between nearest neighbor sites: ε_{AA}, ε_{BB} and ε_{AB}. One can map the binary alloy model into the lattice gas model and the Ising model.

Generalizations to other Spin Models

One can generalize the Ising model in a number of ways, though the Ising spins refer to entities with two possible values. Classically other types of spins are xy spins which can rotate in the x – y plane and have a fixed length (usually $|S| = 1$). They have two components: S_x and S_y. Heisenberg spins are fixed length spins that can point anywhere on a unit sphere. Heisenberg spins have three components: S_x, S_y, and S_z.

Quantum mechanically, the values of S_z are discretized. So Ising spins correspond to $\langle S_z \rangle = \pm 1/2. \vec{S} = \vec{\sigma}/2$ where $\vec{\sigma}$ are the Pauli spin matrices.

We have already mentioned that the Ising model can be considered in higher dimensions. Another variation of the Ising model is to consider other types of lattices. For example in two dimensions, we could have a triangular lattice. In three dimensions there are a wide variety of lattice structures that could be considered. Or one could throw away the lattice and have randomly placed sites to make a type of spin glass.

Other variations center around the form of the interaction. For example, we can allow the nearest neighbor interactions to be antiferromagnetic. Or we can allow the interactions to extend over a longer range to include next nearest-neighbors or even farther, e.g., infinite range. The interaction is contained in the exchange constant J_{ij}. So one could have something like,

$$J_{ij} = \frac{A}{\left|\vec{r}_i - \vec{r}_j\right|^n}$$

where n = 1 is the Coulomb interaction and n = 3 is similar to a dipolar interaction. Another interaction is the RKKY interaction. RKKY stands for Ruderman-Kittel-Kasuya-Yosida. The RKKY interaction has the form:

$$J(r) \sim \frac{\sin(2k_F r) - 2k_F r \cos(2k_F r)}{(k_F r)^4}$$

$$\sim \frac{\cos(2k_F r)}{(k_F r)^3}$$

where k_F is the Fermi wavevector. This interaction is found in metals with magnetic atoms. The interaction is mediated by the conduction electrons. Notice that the interaction oscillates and decays as a power law.

Frustration and Spin Glasses

Magnetic impurities randomly placed in a metal, e.g., Mn impurities in copper, will interact with one another via the RKKY interaction. Because of the oscillations, the interactions will be random. This spin system is called a spin glass. For simplicity, the RKKY interaction is replaced by a random J_{ij} in the spin Hamiltonian:

$$\mathcal{H} = -\sum_{i>j} J_{ij} S_i \cdot S_j$$

Typically, J_{ij} is chosen from a distribution P(J) centered at J = 0. For example, $J_{ij} = \pm J$ where J is a positive constant, and there is an equal probability of choosing the plus or minus sign. Another possibility is to have P(J) be a Gaussian distribution centered at J = 0. Obviously the ground state of a such a system will be disordered, but a phase transition from a paramagnetic phase at high temperatures to a frozen spin configuration is possible.

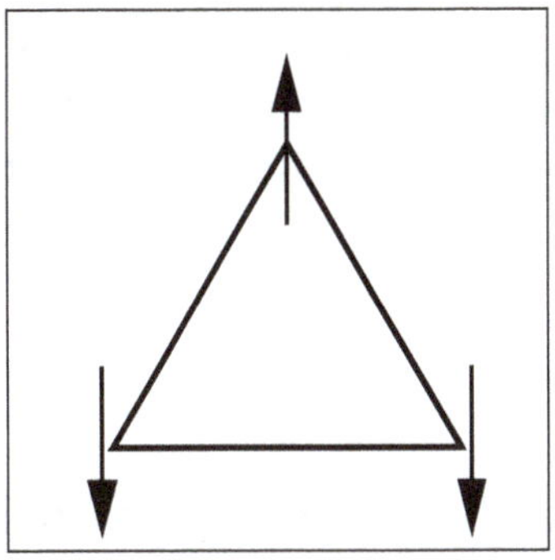

One concept that is associated with spin glasses is "frustration." The idea is best illustrated by considering an equilateral triangle with a spin on each vertex. Suppose the interaction between nearest neighbour spins is antiferromagnetic so that a spin wants to point opposite from its neighbours. There is no way all the spins on the triangle can be satisfied. This is an example of frustration. Frustration occurs when there is no spin configuration where all the spins have their lowest possible interaction energy. In a spin glass, there is a great deal of frustration. As a result there is no clear ground state configuration. One can associate an energy landscape with the energies of different spin configurations. Valleys correspond to low energy configurations and mountains to high energy configurations. The landscape exists in the space of spin configurations. So to go from one valley to another, the system must climb out of the first valley by going through some high energy configurations and then descend into the second valley by passing through configurations with decreasing energy.

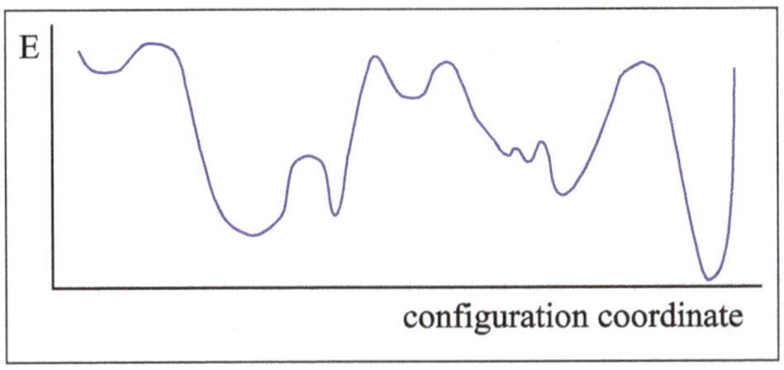

Spin glasses are often used to model interacting systems with randomness. They were originally proposed to explain metals with magnetic impurities and were thought to be a simple model of a glass. Spin glass models have a wide range of applications, e.g., they have been used to model the brain and were the basis of neural networks and models of memory.

Heisenberg Model

The Heisenberg model, developed by Werner Heisenberg, is a statistical mechanical model used in the study of critical points and phase transitions of magnetic systems, in which the spins of the magnetic systems are treated quantum mechanically. It is related to the prototypical Ising model, where at each site of a lattice, a spin $\sigma_i \in \{\pm 1\}$ represents a microscopic magnetic dipole to which the magnetic moment is either up or down. Except the coupling between magnetic dipole moments, there is also a multipolar version of Heisenberg model called the multipolar exchange interaction.

For quantum mechanical reasons, the dominant coupling between two dipoles may cause nearest-neighbors to have lowest energy when they are aligned. Under this assumption (so that magnetic interactions only occur between adjacent dipoles) and on a 1-dimensional periodic lattice, the Hamiltonian can be written in the form:

$$\hat{H} = -J\sum_{j=1}^{N}\sigma_j\sigma_{j+1} - h\sum_{j=1}^{N}\sigma_j,$$

where J is the coupling constant and dipoles are represented by classical vectors (or "spins") σ_j, subject to the periodic boundary condition $\sigma_{N+1} = \sigma_1$.. The Heisenberg model is a more realistic model in that it treats the spins quantum-mechanically, by replacing the spin by a quantum operator acting upon the tensor product $(\mathbb{C}^2)^{\otimes N}$, of dimension 2^N. The Pauli spin-1/2 matrices:

$$\sigma^x = \begin{pmatrix} 0 & 1 \\ 1 & 0 \end{pmatrix},$$

$$\sigma^y = \begin{pmatrix} 0 & -i \\ i & 0 \end{pmatrix},$$

$$\sigma^z = \begin{pmatrix} 1 & 0 \\ 0 & -1 \end{pmatrix},$$

and for $1 \le j \le N$ and $a \in \{x,y,z\}$ denote $\sigma_j^a = I^{\otimes j-1} \otimes \sigma^a \otimes I^{\otimes N-j}$, where I is the 2×2 identity matrix. Given a choice of real-valued coupling constants J_x, J_y, and, the Hamiltonian is given by,

$$\hat{H} = -\frac{1}{2}\sum_{j=1}^{N}\left(J_x\sigma_j^x\sigma_{j+1}^x + J_y\sigma_j^y\sigma_{j+1}^y + J_z\sigma_j^z\sigma_{j+1}^z + h\sigma_j^z\right)$$

where the h on the right-hand side indicates the external magnetic field, with periodic boundary conditions. The objective is to determine the spectrum of the Hamiltonian, from which the partition function can be calculated and the thermodynamics of the system can be studied.

It is common to name the model depending on the values of J_x, J_y, and J_z: if $J_x \ne J_y \ne J_z$, the model is called the Heisenberg XYZ model; in the case of $J = J_x = J_y \ne J_z = \Delta$,, it is the Heisenberg XXZ model; if $J_x = J_y = J_z = J$, it is the Heisenberg XXX model. The spin 1/2 Heisenberg model in one dimension may be solved exactly using the Bethe ansatz. In the algebraic formulation, these are related to particular quantum affine algebras and elliptic quantum groups in the XXZ and XYZ cases respectively. Other approaches do so without Bethe ansatz.

XXX Model

The physics of the Heisenberg XXX model strongly depends on the sign of the coupling constant J and the dimension of the space. For positive J the ground state is always ferromagnetic. At negative

J the ground state is antiferromagnetic in two and three dimensions. In one dimension the nature of correlations in the antiferromagnetic Heisenberg model depends on the spin of the magnetic dipoles. If the spin is integer then only short-range order is present. A system of half-integer spins exhibits quasi-long range order.

A simplified version of Heisenberg model is the one-dimensional Ising model, where the transverse magnetic field is in the *x*-direction, and the interaction is only in the *z*-direction:

$$\hat{H} = -J\sum_{j=1}^{N} \sigma_j^z \sigma_{j+1}^z - gJ\sum_{j=1}^{N} \sigma_j^x.$$

At small g and large g, the ground state degeneracy is different, which implies that there must be a quantum phase transition in between. It can be solved exactly for the critical point using the duality analysis. The duality transition of the Pauli matrices is $\sigma_i^z = \prod_{j \le i} S_j^x$ and $\sigma_i^x = S_i^z S_{i+1}^z$, where S^x and S^z are also Pauli matrices which obey the Pauli matrix algebra. Under periodic boundary conditions, the transformed Hamiltonian can be shown is of a very similar form:

$$\hat{H} = -gJ\sum_{j=1}^{N} S_j^z S_{j+1}^z - J\sum_{j=1}^{N} S_j^x$$

but for the g attached to the spin interaction term. Assuming that there's only one critical point, we can conclude that the phase transition happens at g = 1.

Applications

- Another important object is entanglement entropy. One way to describe it is to subdivide the unique ground state into a block (several sequential spins) and the environment (the rest of the ground state). The entropy of the block can be considered as entanglement entropy. At zero temperature in the critical region (thermodynamic limit) it scales logarithmically with the size of the block. As the temperature increases the logarithmic dependence changes into a linear function. For large temperatures linear dependence follows from the second law of thermodynamics.
- The Heisenberg model provides an important and tractable theoretical example for applying density matrix renormalisation.
- The six-vertex model can be solved using the algebraic Bethe ansatz for the Heisenberg spin chain.
- The half-filled Hubbard model in the limit of strong repulsive interactions can be mapped onto a Heisenberg model with $J < 0$ representing the strength of the super exchange interaction.

Percolation Theory

Percolation theory is the simplest not exactly solved model displaying a phase transition. Often, the insight into the percolation theory problem facilitates the understanding of many other physical systems. Moreover, the concept of fractals, which is intimately related to the percolation theory problem, is of general interest as it pops up more or less everywhere in Nature. The knowledge of percolation, fractals, and scaling are of immense importance theoretically in such diverse fields as biology, physics, and geophysics and also of practical importance in e.g. oil recovery. We will begin gently by developing a basic understanding of percolation theory, providing a natural introduction to the concept of scaling and renormalisation group theory.

Preliminaries

Let P(A) denote the probability for an event A and $P(A_1 \cap A_2)$ the joint probability for event A_1 and A_2.

Definition 1: Two events A_1 and A_2 are independent $\Leftrightarrow P(A_1 \cap A_2) = P(A_1)P(A_2)$.

Definition 2: More generally, we define $n \geq 3$ events $A_1, A_2, \ldots, A_n$ to be mutually independent if $P(A_1 \cap A_2 \cap \cdots \cap A_n) = P(A_1)P(A_2)\cdots P(A_n)$ and if any subcollection containing at least two but fewer than n events are mutually independent.

Let each site in a lattice be occupied at random with probability p, that is, each site is occupied (with probability p) or empty (with probability 1−p) independent of the status (empty or occupied) of any of the other sites in the lattice. We call p the occupation probability or the concentration.

Definition 3: A cluster is a group of nearest neighbouring occupied sites.

Percolation theory deals with the numbers and properties of the clusters formed when sites are occupied with probability p.

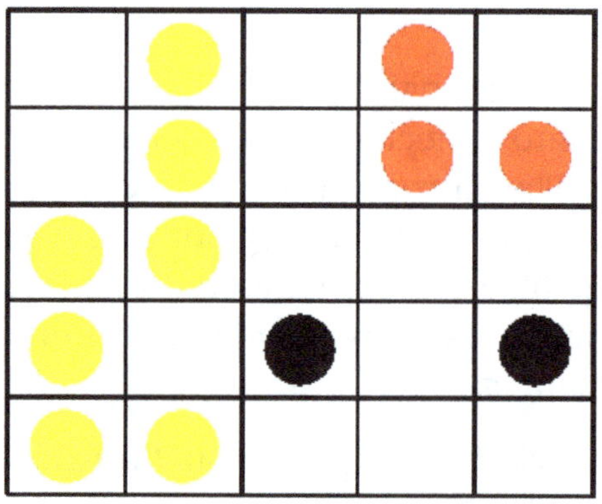

Percolation in 2d square lattice of linear size L = 5. Sites are occupied with probability p. In the lattice above, we have one cluster of size 7, a cluster of size 3 and two clusters of size 1 (isolated sites).

Definition 4: The cluster number $n_s(p)$ denotes the number of s-clusters per lattice site.

The (average) number of clusters of size s in a hypercubic lattice of linear size L is $L^d n_s(p)$, d being the dimensionality of the lattice. Defining the cluster number per lattice site as opposed to the total number of s-clusters in the lattice ensures that the quantity will be independent of the lattice size L.

For finite lattices $L < \infty$, it is intuitively clear, that if the occupation probability p is small, there is only a very tiny chance of having a cluster percolating between two opposite boundaries (i.e., in 2d, from top-to-bottom or from left-to-right). For p close to 1, we almost certainly will have a cluster percolating through the system. In figure below, sites in 2d square lattices are occupied at random with increasing occupation probability p. The occupied sites are shown in gray while the sites belonging to the largest cluster are shown in black. Unoccupied sites are white. Note that for $p \approx 0.59$, a percolating cluster appears for the first time.

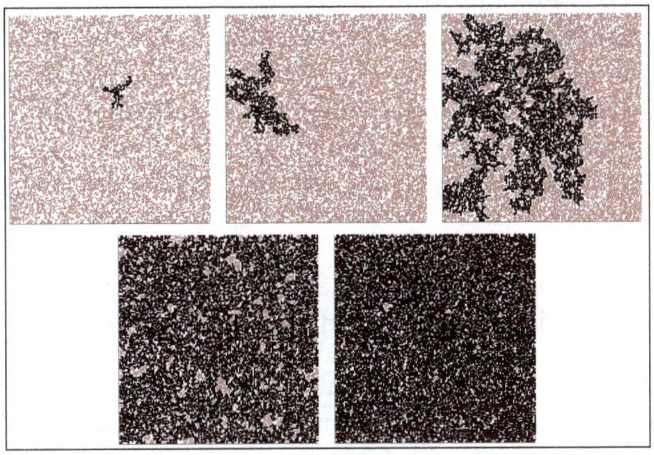

Percolation in 2d square lattices with system size $L \times L = 150 \times 150$. Occupation probability $p = 0.45, 0.55, 0.59, 0.65$, and 0.75, respectively. Notice, that the largest cluster percolates through the lattice from top to bottom in this example when $p \geq 0.59$.

Definition 5: The percolation threshold p_c is the concentration (occupation probability) p at which an infinite cluster appears for the first time in an infinite lattice. p_c is defined with respect to an infinite lattice, that is, in the limit of $L \to \infty$. Table lists the percolation threshold in various lattices and dimensions.

Percolation in 1d

We will consider the percolation problem in 1d where it can be solved analytically. Many of the characteristic features encountered in higher dimensions are present in 1d as well, if we know

Lattice	# nn	Site percolation	Bond percolation
1d	2	1	1
2d Honeycomb	3	0.6962	$1 - 2\sin(\pi/18) \approx 0.65271$
2d Square	4	0.592746	1/2
2d Triangular	6	1/2	$2\sin(\pi/18) \approx 0.34729$
3d Diamond	4	0.43	0.388
3d Simple cubic	6	0.3116	0.2488
3d BCC	8	0.246	0.1803
3d FCC	12	0.198	0.119
4d Hypercubic	8	0.197	0.1601
5d Hypercubic	10	0.141	0.1182
6d Hypercubic	12	0.107	0.0942
7d Hypercubic	14	0.089	0.0787
Bethe lattice	z	1/(z-1)	1/(z-1)

Table shows the percolation threshold for the site percolation problem is given in column 3 for various lattices in various dimensions. Column 2 lists the number of nearest-neighbours (nn), also known as the coordination number. Within a given dimension, the percolation threshold decrease with increasing number of nearest-neighbours. The site percolation problem has a counterpart called the bond percolation problem: In a lattice, each bond between neighbouring lattice sites can be occupied (open) with probability p and empty (closed) with probability (1−p). A cluster is a group of connected occupied (open) bonds. NB: In all cases, a cluster is defined as a group of nearest neighbouring occupied sites (bonds). Note that the percolation threshold for the site-percolation on high-dimensional hypercubic lattices, where loops become irrelevant, approaches that of the Bethe lattice $1/(z-1)$, if we substitute the coordination number z with 2d.

Where and how to look. Thus the 1d case serves as a transparent window into the world of phase transitions, scaling, scaling relations, and renormalisation group theory.

Imagine a 1d lattice with an infinite number of sites of equal spacing arranged in a line. Each site has a probability p of being occupied, and consequently $1-p$ of being empty (not occupied). These are the only two states possible as shown in figure.

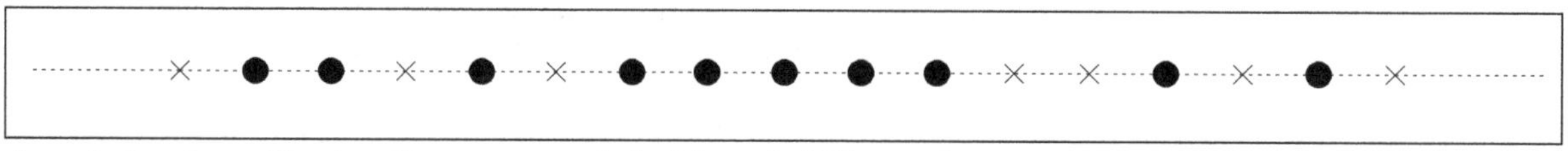

Percolation in a 1d lattice. Sites are occupied with probability p. The crosses are empty sites, the solid circles are occupied sites. In the part of the infinite 1d lattice shown above, there is one cluster of size 5, one cluster of size 2, and three clusters of size 1.

What is of interest to us now and in future discussions in higher dimensions is the occupation probability at which an infinite cluster is obtained for the first time. A percolating cluster in 1d spans from $-\infty$ to $+\infty$. Clearly, in 1d this can only be achieved if all sites are occupied, that is, the percolation threshold $p_c = 1$, as a single empty site would prevent a cluster to percolate. A precise "mathematical" derivation of $p_c = 1$ in 1d goes as follows:

Definition: Let $\Pi(p,L)$ denote the probability that a lattice of linear size L percolates at concentration p.

Combining the two definitions we have,

$$\lim_{L\to\infty} \Pi(p,L) = \begin{cases} 0 & \text{for } p < p_c \\ 1 & \text{for } p \geq p_c. \end{cases}$$

Consider a finite 1d lattice of size L where each site is occupied with probability p. As the events of occupying sites are independent, all sites are occupied with probability $\Pi(p,L) = p^L$ and,

$$\lim_{L\to\infty} \Pi(p,L) = \lim_{L\to\infty} p^L = \begin{cases} 0 & \text{for } p < 1 \\ 1 & \text{for } p = 1, \end{cases}$$

implying $p_c = 1$.

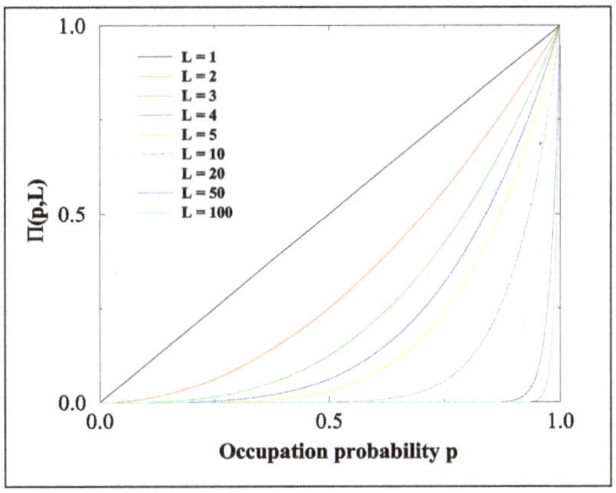

The probability of a 1d lattice of linear size L to percolation at occupation probability p. In the limit $L \to \infty$, $\Pi(p,L)$ converges to a discontinous step function.

Let us consider the clusters formed in a 1d lattice. A cluster of size s is formed when s sites are occupied next to one another bounded by two empty sites, as shown in figure above. When $L \to \infty$, we can ignore the effects of the boundary sites of the lattice and the probability of an arbitrary site (occupied or not) being, say, the left hand side (LHS) of an s-cluster is,

$$n_s(p) = (1-p)p^s(1-p) = (1-p)^2 p_s.$$

This expression is obtained from the assumption that the occupancy of each site is independent of the state of any other site. If this was not the case then it would be much more complicated. Note that, since all sites have equal probability of being occupied (or empty), the probability that an arbitrary site is part of an s-cluster is s times the probability of it being the LHS of the cluster.

We can re-write the cluster number Equation $n_s(p) = (1-p)p^s(1-p) = (1-p)^2 p_s$ for 1d percolation as:

$$n_s(p) = (1-p)^2 p_s$$

$$= (1-p)^2 \exp\left(\text{In}\left(p^s\right)\right)$$

$$= (1-p)^2 \exp\left(s \,\text{In}(p)\right)$$

$$= (pc-p)^2 \exp\left(-\frac{s}{s_\xi}\right)$$

with the definition of a cutoff cluster size or characteristic cluster size,

$$s_\xi = \frac{-1}{\text{In}(p)} = \frac{-1}{\text{In}(p_c-(p_c-p))} \to \frac{1}{p_c-p} = (p_c-p)^{-1} \text{ for } p \to p_c,$$

where we, to obtain the limit, have used $p_c = 1$ and the Taylor expansion:

$$\text{In}(1-x) = -x - \frac{1}{2}x^2 - \frac{1}{3}x^3 - \ldots \approx -x$$

where the last approximation is valid for $x \to 0$:

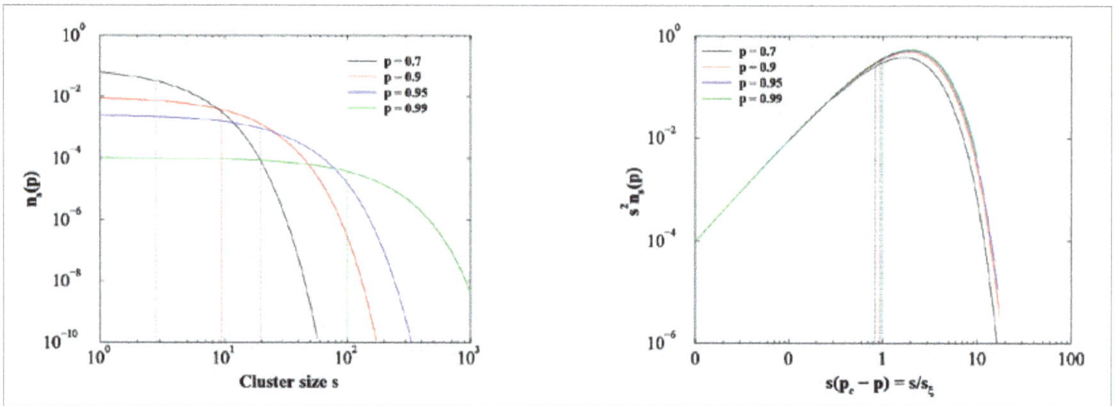

Percolation in 1d. (a) The cluster number distribution $n_s(p) = (p_c - p)^2 \exp\left(-\frac{s}{s_\xi}\right)$ for various values of p approaching $p_c = 1$. The vertical lines indicate the cutoff cluster size $s_\xi(p)$. (b) By plotting $s^2 n_s(p)$ versus $s/s_\xi \approx s(p_c - p)$, all the data collapses onto a function $x^2 \exp(-x)$. (c) The characteristic cluster size $s_\xi = -\frac{1}{\ln(p)}$ diverges when $p \to p_c = 1$. (d) In the limit $p \to p_c = 1$, $s_\xi \propto (p_c - p)^{-1}$.

Example: Verify the Taylor expansion of ln(1 – x) around x = 0 given above.

Thus the cutoff cluster size s_ξ diverges for $p \to p_c$ as a power law in the distance from the critical occupation probability pc, as shown in figure above. The divergence of the cutoff cluster size when $p \to pc$ is also seen in higher dimensions where, however, another numerical exponent will describe the divergence. Thus it is natural to introduce a symbol for the exponent.

Definition: The critical exponent σ is defined by,

$$s_\xi \propto |p_c - p|^{-\frac{1}{\sigma}} \text{ for } p \to p_c.$$

In 1d percolation theory, σ = 1 and log s_ξ = constant $-\frac{1}{\sigma}\log|p_c - p|$,

Let us continue our journey into 1d percolation theory. For $p < p_c$ we can state that the probability that an arbitrary site belongs to any (finite) cluster is simply the probability p of it being occupied. Since the probability that an arbitrary sites belongs to an s-cluster is given by $sn_s(p)$, we arrive at:

$$\sum_{s=1}^{\infty} sn_s(p) = p \qquad \text{for } p < p_c.$$

Using the formula for summing a geometric series, we can satisfy those who prefer a rigorous mathematical proof:

$$\sum_{s=1}^{\infty} sn_s(p) = \sum_{s=1}^{\infty} s(1-p)^2 p^s$$

$$=(1-p)^2\sum_{s=1}^{\infty}p\frac{d(p^s)}{dp}$$

$$=(1-p)^2 p\frac{d}{dp}\left(\sum_{s=1}^{\infty}p^s\right)$$

$$=(1-p)^2 p\frac{d}{dp}\left(\frac{p}{1-p}\right)$$

$$=p.$$

How large on average is a cluster or, equivalently, how large is a cluster on average to which an occupied site belongs? The probability that a site is occupied is p. The probability that an arbitrary site belongs to an s-cluster is $sn_s(p)$. Thus the probability w_s that the cluster to which an occupied sites belongs contains s sites is,

$$w_s=\frac{sn_s(p)}{p}=\frac{sn_s(p)}{\sum_{s=1}^{\infty}sn_s(p)}.$$

Thus, the mean cluster size or average cluster size S(p) is given by,

$$S(p)=\sum_{s=1}^{\infty}sw_s$$

$$=\sum_{s=1}^{\infty}\frac{s^2n_s(p)}{\sum_{s=1}^{\infty}n_s(p)s}$$

$$=\frac{1}{p}(1-p)^2\sum_{s=1}^{\infty}s^2p^2$$

$$=\frac{1}{p}(1-p)^2\left(p\frac{d}{dp}\right)^2\left(\sum_{s=1}^{\infty}p^s\right),$$

where the operator,

$$\left(p\frac{d}{dp}\right)^2=\left(p\frac{d}{dp}\right)\left(p\frac{d}{dp}\right)\neq p^2\frac{d^2}{dp^2}.$$

Using the formula for summing a geometric series and the operator $p\frac{d}{dp}$ twice, we finally arrive at: $S(p)=\frac{1+p}{1-p}=\frac{p_c+p}{p_c-p}$

where the last equality follows because in 1d the critical occupation probability $p_c = 1$.

Example: Derive the result in Equation $S(p)=\frac{1+p}{1-p}=\frac{p_c+p}{p_c-p}$ for the mean cluster size in 1d.

We thus see that the mean cluster size diverges for $p\rightarrow p_c^-$, where the minus sign signifies that we are approaching pc from below, which is what we intuitively expect if considering an infinite lattice. It is not possible to approach pc from above in 1d as $p_c = 1$. This is actually the main difference between 1d and higher dimensions, where, $p_c < 1$ and we can approach p_c from both below and above.

In order to investigate in detail how the mean cluster size diverges, when taking the limit $p \to p_c^-$,, we note that the numerator in Equation $S(p) = \frac{1+p}{1-p} = \frac{p_c + p}{p_c - p}$ approaches $2p_c$, so,

$$S(p) = \frac{p_c + p}{p_c - p} \to \frac{2p_c}{p_c - p} \propto (p_c - p)^{-1} \text{ for } p \to p_c^-$$

Thus in 1d, the mean cluster diverges like a power law in the quantity $(p_c - p)$ when $p \to p_c$, as shown in figure below. The same phenomenon will be encountered in higher dimensions.

Definition: The critical exponent γ is defined by,

$$S(p) \propto |p_c - p|^{-\gamma} \text{ for } p \to p_c.$$

In 1d percolation theory, $\gamma = 1$.

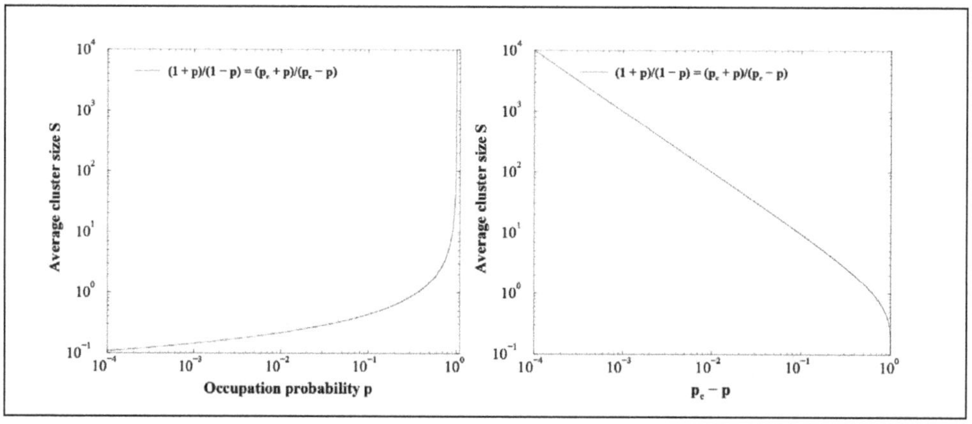

Percolation in 1d. The average cluster size $S(p) = (1 + p)/(1 - p) = (p_c + p)/(p_c - p)$ diverges when $p \to p_c = 1$. In the limit $p \to p_c = 1$, $S(p) \to 2/(p_c - p)$.

Definition: The correlation function or pair connectivity g(r) is the probability that a site at position r from an occupied site belongs to the same finite cluster.

Note this definition excludes the contribution from the infinite cluster. That need not worry us in 1d, where all clusters are finite if $p < p_c = 1$. Let $r = |r|$. Clearly, $g(r = 0) = 1$, since the site is occupied by definition. In 1d, for a site at position r to be occupied and belong to the same (finite) cluster, this site and the $(r - 1)$ intermediate sites must be occupied, leaving,

$$g(r) = p^r,$$

for all p, which can also be expressed in the form,

$$g(r) = \exp\left(\text{In}(p^r)\right) = \exp\left(r\,\text{In}(p)\right) = \exp\left(-\frac{r}{\xi}\right),$$

Where,

$$\xi = -\frac{1}{\ln(p)} = \frac{-1}{\ln(p_c - p)} \to \frac{1}{(p_c - p)} = (p_c - p)^{-1} \quad p \to p_c = 1,$$

where we use the expansion $\ln(1-x) \approx -x$ for small x. The quantity ξ is called the correlation length which diverges for $p \to p_c$. The same phenomenon will be encountered in higher dimensions. Note that in 1d we have $s_\xi = \xi$ which is why we haven't bothered displaying a figure for ξ, as it would be identical to figure. However, this identity will not be true in higher dimensions, where we will find $s_\xi \propto \xi^D$, where D is the fractal dimension.

Definition: The critical exponent v is defined by,

$$\xi \propto |p_c - p|^{-v} \text{ for } p \to p_c.$$

In 1d percolation theory, v = 1.

By summing over all possible lattice sites r of the correlation function, the mean cluster size can be shown to be,

$$\sum_r g(r) = S(p).$$

In 1d, this sum is straight forward identifying r with r = 0, ±1, ±2,..., the general pattern for the exact solutions of the 1d percolation problem is that certain quantities, such as the cutoff cluster size s_ξ, the mean cluster size S(p), and the correlation length ξ diverge at the percolation threshold. The divergence can be described by simple power laws of the distance from the critical occupation probability $|\Delta p| = |p_c - p|$, e.g. $\xi \propto (p_c - p)^{-1}$,, at least asymptotically close to p_c where Δp is small.

Percolation in the Bethe Lattice

The percolation problem can be solved analytically in d = 1 and d = ∞. The infinitely dimensional case is synonymous with the Bethe lattice, a special lattice where each site has z neighbouring sites, such that each branch gives rise to z – 1 other branches, as shown in figure below.

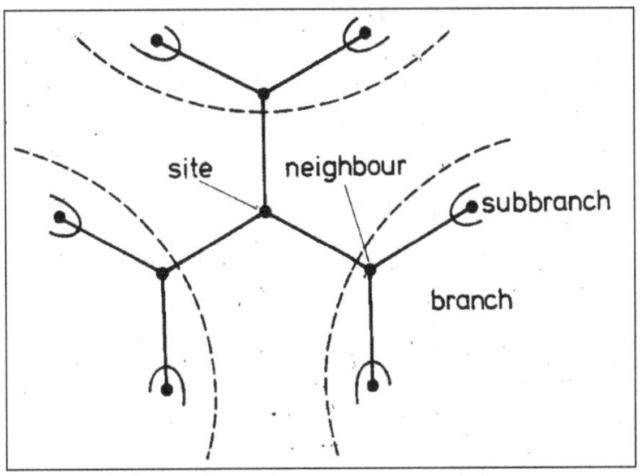

The Bethe lattice with z = 3. Each site has three neighbours. Each branch contains z – 1 = 2 subbranches.

The 1d case is effectively a Bethe lattice with z = 2. Why does the Bethe lattice correspond to the spatial dimension $d = \infty$ you might rightly ask! Well, in a hypercubic lattice, (a) the number of surface sites relative to the total number of sites approaches a constant when $d \to \infty$ and (b) there are no closed loops when $d \to \infty$. The Bethe lattice has both these properties.

Let g denote the generation, that is, the distance from a "centre site". However, that in an infinite Bethe lattice, all sites are equivalent, so the notion of a "centre site" is not to be taken literally. In the figure above, the first "ring" of three sites belong to generation g = 1, the second "ring" of six sites belongs to the second generation and so on. The total number of sites in a Bethe lattice consisting of g generations is,

$$\begin{aligned}\text{Total no. sites} &= 1+3\cdot\left(1+2+\cdots+2^{g-1}\right)\\ &= 1+3\cdot\frac{1-2^g}{1-2}\\ &= 3\cdot 2^g-2,\end{aligned}$$

while the number of surface sites is $3 \cdot 2^{g-1}$. Thus,

$$\frac{\text{No. of surface sites}}{\text{Total no. of sites}} = \frac{3\cdot 2^{g-1}}{3\cdot 2^g-2} \to \frac{1}{2} \quad \text{for } g \to \infty$$

and the surface/volume tends to a constant.

Example: Show, that for a general Bethe lattice with coordination number z:

$$\frac{\text{No. of surface sites}}{\text{Total no. of sites}} \to \frac{z-2}{z-1} \text{for } g \to \infty$$

In a hypercubic lattice of linear size L, the surface is proportional to the volume only when $d \to \infty$: the surface in d dimensions is proportional to L^{d-1} while the volume is proportional to L^d leaving,

$$\text{Surface} \propto \text{Volume}^{\frac{d-1}{d}} = \text{Volume}^{1-\frac{1}{d}},$$

that is, the surface is proportional to the volume if $d \to \infty$.

There are no closed loops in a Bethe lattice. Starting from the "centre site" going outwards, one will never return to the starting point. In a hypercubic lattice, the chance (probability) of having a loop approaches zero as the dimension $d \to \infty$: As an example, let us place four particles in a chain in a hypercubic lattice with dimension d. When the first particle has been placed, there are 2d nearest neighbour sites, where the second particle can be placed. However, for the third and fourth particle, there are only 2d – 1 possible sites, implying a total no. of different chains $2d \cdot (2d-1)^2$. Calculating the number of ways to place four particles in a loop, we arrive at $2d \cdot (2d-2) \cdot 1$, that is, the probability of having a four loop:

$$\frac{\text{No. loops}}{\text{Total no. chains}} = \frac{2d\cdot(2d-2)\cdot 1}{2d\cdot(2d-1)^2} = \frac{(2d-2)}{(2d-1)} \to 0 \quad \text{for } d \to \infty.$$

For $d = \infty$, there is no chance of having a closed loop, which is intuitively clear, isn't it?

What is the critical occupation probability in a Bethe lattice, that is, at which occupation probability does an infinite cluster (path) appear for the first time in an infinite Bethe lattice? Starting from a "centre site" and going outwards, we encounter (z – 1) new neighbours. Thus, on average, we have p(z – 1) new occupied sites on which we can continue the path. The critical occupation probability is determined by the equation,

$$p_c(z-1)=1 \Leftrightarrow p_c = \frac{1}{z-1}.$$

Notice, that for z = 2, the 1d percolation problem, we recover pc = 1. For z = 3, e.g., $p_c = \frac{1}{2}$. Note that in a hypercubic lattice with $\rightarrow\infty$ one would expect $p_c \rightarrow 1/(2d-1)$ in order to be able to continue walking along a path of occupied sites.

Definition: The strength of the infinite cluster P(p) is the probability that an arbitrary site belongs to the infinite cluster.

Example: Discuss why, obviously, P(p) = 0 for $p < p_c$ in an infinite lattice.

In order to calculate P(p) for p > pc in the Bethe lattice, we introduce the quantity Q as the probability that an arbitrary site is NOT connected to infinity through a fixed branch originating at this site. Restricting ourselves to the lattice with z = 3 and using basic probability theory, we have that the strength P(p) of the infinite network, that is, the probability that an arbitrarily selected site is connected to infinity by occupied sites is,

$$\begin{aligned} P(p) &= (\text{Prob. site is occupied}) \times (\text{Prob. at least ONE branch lead to infinity}) \\ &= p(1 - Q^3). \end{aligned}$$

Next, we need to determine Q, the probability that a branch originating at that site does not lead to infinity. This depends on whether or not the neighbouring site is occupied. We find,

$$\begin{aligned} Q &= (\text{Prob. site is empty}) + (\text{Prob. site is occupied}) \times (\text{Prob. no subbranch leads to infinity}) \\ &= (1 - p) + pQ^2. \end{aligned}$$

We have relied on the fact that all sites in a Bethe lattice are equivalent, so Q is also the probability that a subbranch is not connected to infinity. The quadratic equation can easily be solved:

$$Q = \frac{1 \pm \sqrt{(2p-1)^2}}{2p} = \begin{cases} 1 \\ \frac{1-p}{p} \end{cases}.$$

Example: Show that, in a general Bethe lattice where each site has z neighbours,

$$Q = \begin{cases} 1 \\ 1 - \frac{2p(z-1)-2}{p(z-1)(z-2)} \end{cases}.$$

When $p < p_c$, there are no infinite clusters, by definition, so with probability 1 there is no connection to infinity. Thus Q = 1 is the trivial solution associated with P(p) = 0 for $p < p_c$. The non-trivial solution $Q = \frac{1-p}{p}$ is associated with $p > p_c$, leaving,

$$P(p) = \begin{cases} 0 & \text{for } p < p_c \\ p\left(1 - \left(\frac{1-p}{p}\right)^3\right) & \text{for } p \geq pc. \end{cases}$$

Using a Taylor expansion for P(p) around $p = p_c = \frac{1}{2}$, it can be shown that,

$$P(p) \propto (p - p_c) \text{ for } p \to p_c^+.$$

Definition: The critical exponent β is defined by,

$$P(p) \propto (p - p_c)^\beta \text{ for } p \to p_c^+.$$

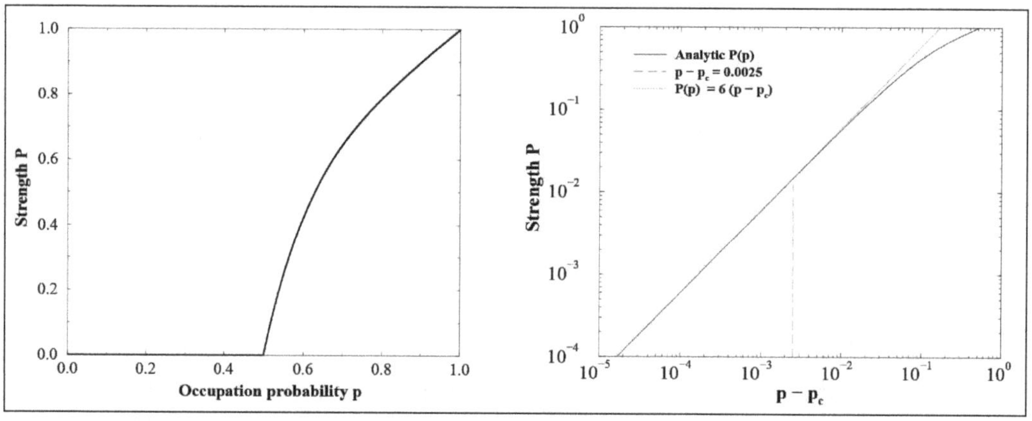

The strength P(p) of the infinite cluster in a Bethe lattice with z = 3. When $p \to p_c^+$, the strength $P(p) \to 6(p - p_c)$ implying β = 1 in a Bethe lattice.

In the Bethe lattice, β = 1 as shown in figure above. The strength P(p) measures how large a fraction of the sites in the lattice belongs to the infinite cluster (in an infinite lattice) and is known as the order parameter. The phenomenon that the order parameter (strength) becomes non-zero for $p > p_c$ is known as a phase transition and p = pc is often called a critical point or critical occupation probability.

Example: What is the strength P(p) in the 1d percolation problem?

What is the mean cluster size S(p) to which an occupied site (e.g. the "centre site") belongs? Let T be the mean cluster size in one branch.

S(p) = average cluster size to which the origin belongs

= (contribution from origin) + (contributions from the 3 branches)

= 1 + 3T.

Thus, we have to determine T. Again, as all sites are equivalent in an infinite Bethe lattice, T will also denote the mean cluster size in the subbranch, so we can argue as follows: If the neighbour to the "centre site" is empty, probability (1 – p), there is no contribution to T from this branch. If the neighbour to the "centre site" is occupied (probability p), it contributes its own mass (unity) to the cluster, and adds the mass T for each of its z – 1 = 2 subbranches. Thus,

$$T=(1-p)\cdot 0+p\cdot(1+2T)\Leftrightarrow T=\frac{p}{1-2p}\quad \text{for } p<p_c$$

Therefore, the mean cluster size S(p) is given by,

$$S(p)=1+3T=\frac{1+p}{(1-2p)}=\frac{1+p}{2\left(\frac{1}{2}-p\right)}=\frac{1+p}{2(p_c-p)}=\Gamma_2(p_c-p)^{-1},$$

Where, $\Gamma_2=\frac{1+p}{2}$ is the amplitude, and the critical exponent $\gamma=1$ in the Bethe lattice. This is the exact result for a mean cluster of size S(p) and we notice that it diverges for $p\to p_c$.

An occupied site either belongs to the infinite cluster or a finite cluster. Thus, the generalisation of Equation $\sum_{s=1}^{\infty} sn_s(p)=p$ for $p<p_c$. is valid for all p,

$$P(p)+\sum_{s=1}^{\infty} sn_s(p)=p\quad \forall p.$$

The sum runs over all finite s and excludes the infinite cluster. Of course, Equation above reduces to Equation $\sum_{s=1}^{\infty} sn_s(p)=p$ for $p<p_c$. for p < pc since the strength P(p) is identical zero in this range.

Example: Reflect upon the identity Equation $P(p)+\sum_{s=1}^{\infty} sn_s(p)=p\quad \forall p$. for p = p_c. What is the sum $\sum_{s=1}^{\infty} sn_s(p_c)$? Is your answer also valid in d = 1?

Critical phenomena occur in a variety of systems. The Bethe lattice approximation for percolation theory is somewhat analogous to the mean-field approximation for magnetic systems dislaying a phase transition, or the Van der Waals equation for a fluid. In all cases, the order parameter becomes non-zero above (or below) the critical point. In a magnetic system, the order parameter is the spontaneous magnetisation per spin m, which is zero above the critical temperature (Curie temperature) T_c and non-zero for $T < T_c$ as the system enters an ordered state. We will see later, that $m \propto (T_c - T)^{\beta}$. In the case of a fluid, the order parameter is identified as the the difference between liquid and vapour density which becomes nonzero as $T < T_c$. The analogy between meanfield thermal critical phenomena and percolation is not complete since the critical exponent β for the order parameter is $\frac{1}{2}$ for thermal phase transitions and unity for percolation, but nevertheless, the qualitative similarities are clear.

In both the thermal phase transitions mentioned above and percolation, the order parameter goes to zero continuously as one approaches the critical point. Such phase transitions are called continuous phase transitions or second-order phase transitions. If instead the order parameter jumps to zero, one has a first order phase transition. Such transitions can occur in more complicated percolation theory situations, like "bootstrap percolation" where on a square lattice a site remains occupied only if three or four of its neighbours are still occupied.

Cluster Number Distribution

Unfortunately, it is not possible to obtain an exact form for the cluster number $n_s(p)$ in $d > 1$ or the Bethe lattice since, unlike in 1d, there are a very large number of different ways in which clusters can arrange themselves. Even for relatively small cluster numbers in 2d we run into difficulties. Exact enumeration of small cluster "animals" has been tabulated. In a 2d square lattice a computer has counted a total of 68, 557, 762, 666, 345, 165, 410, 168, 738 different cluster configurations with s = 46. Nevertheless, we are still able to account for the general scaling behaviours of certain quantities that characterise the clusters.

Definition: The perimeter or external boundary t of a cluster is the number of empty nearest neighbours of the cluster.

Using the concept of a perimeter, we are able write down the general expression for the cluster number $n_s(p)$ as:

$$n_s(p) = \sum_t g_{s,t} p^s (1-p)^t,$$

where we have introduced $g_{s,t}$, the number of different lattice configurations with size s and perimeter t.

In 1d, all clusters have two perimeter sites and thus,

$$g_{s,t} = \begin{cases} 1 & \text{for } t = 2 \\ 0 & \text{otherwise} \end{cases}$$

implying $n_s(p) = p^s(1 - p)^2$.

Example: Consider clusters of size s = 3 in 2d on a square lattice. Verify, that $g_{s,t}$ is only non-zero for t = 7 or t = 8 and that $g_{s=3,t=7} = 4$ and $g_{s=3,t=8} = 2$, implying $n_{s=3}(p) = 4p^3(1-p)^7 + 2p^3(1-p)^8$.

In a Bethe lattice, however, we can show that there is a unique perimeter for a given cluster size s.

Example: Show that in a Bethe lattice $t = 2 + s(z - 2)$.

Now we apply this formula to our general result for cluster numbers $n_s(p) = \sum_t g_{s,t} p^s (1-p)^t$, to get:

$$n_s(p) = \sum_t g_{s,t} p^s (1-p)^t = g_{s,2+s(z-2)} p^s (1-p)^{2+s(z-2)}$$

For simplicity, from here on we shall work with the z = 3 case rather than the general Bethe lattice.

For $z = 3$ we have $p_c = \frac{1}{2}$. As we have said, determining $g_{s,2+s}$ for large s is difficult, but we can avoid having to calculate $g_{s,2+s}$ by considering the ratio,

$$\frac{n_s(p)}{n_s(p_c)} = \left[\frac{(1-p)}{(1-p_c)}\right]^2 \left[\left(\frac{p}{p_c}\right)\frac{(1-p)}{(1-p_c)}\right]^s$$
$$= \left[\frac{(1-p)}{(1-p_c)}\right]^2 \exp\left[s\,\text{In}\left(\frac{p}{p_c}\frac{(1-p)}{(1-p_c)}\right)\right].$$

In order to proceed, define the function:

$$f(p) = \frac{p}{p_c}\frac{(1-p)}{(1-p_c)} = \frac{p}{\frac{1}{2}}\frac{(1-p)}{\left(1-\frac{1}{2}\right)} = 4p - 4p^2 \quad \text{for } z = 3.$$

Example: Make a Taylor expansion of the function $f(p) = 4p - 4p^2$ around $p = \frac{1}{2} = p_c$ and verify $f(p) = 1 - 4(p - p_c)^2$.

Substituting the expression for f(p) given in Example above, we find,

$$\frac{n_s(p)}{n_s(p_c)} = \left[\frac{(1-p)}{(1-p_c)}\right]^2 \exp\left(-\frac{s}{s_\xi}\right)$$

with,

$$s_\xi = \frac{-1}{\text{In}\left(1-4\left(p-\frac{1}{2}\right)^2\right)}$$
$$\propto \frac{1}{4}\left(p-\frac{1}{2}\right)^{-2} \quad \text{for } p \to \frac{1}{2}$$
$$\propto |p - p_c|^{-\frac{1}{\sigma}}.$$

In the Bethe lattice, $\sigma = \frac{1}{2}$.

We now have a very simple exponential decay for the ratio of cluster numbers. This is to say that large clusters with $s \gg s_\xi$ are very rare indeed. This explains why s_ξ is known as the cutoff cluster size. In order to investigate the cluster number distribution in more detail, we have to consider the term $n_s(pc)$ appearing in the denominator.

From the analysis on the average cluster size, we know that,

$$S(p) = \sum_{s=1}^{\infty} \frac{s^2 n_s(p)}{\sum_{s=1}^{\infty} s n_s(p)} \to \infty \quad \text{for } p \to p_c.$$

However, the denominator remains finite at p = pc (why?) so the numerator must diverge for p = p_c, that is,

$$\sum_{s=1}^{\infty} s^2 n_s(p_c) = \infty.$$

Generally, we have that,

$$\sum_{s=1}^{\infty} s^a = \begin{cases} \text{convergent} & \text{for } a < -1 \\ \text{divergent} & \text{for } a \geq -1. \end{cases}$$

Thus a power-law decay,

$$n_s(p_c) \propto s^{-T} \Rightarrow \sum_{s=1}^{\infty} s^2 n_s(p_c) = \sum_{s=1}^{\infty} s^{2-\tau}$$

would imply a divergence of the average cluster size if the critical exponent $\tau \leq 3$.

However, see Equation $P(p) + \sum_{s=1}^{\infty} sn_s(p) = p \quad \forall p$. We also need $p_c = \sum_{s=1}^{\infty} sn_s(p_c) \propto \sum_{s=1}^{\infty} s^{1-\tau} < \infty$ implying that $\tau > 2$. Therefore, we arrive at a general form for the cluster number distribution in the Bethe lattice:

$$n_s(p) \propto s^{-\tau} \exp\left(-\frac{s}{s_\xi}\right), \quad \text{for } s \gg 1$$

with,

$$s_\xi \propto |p_c - p|^{-\frac{1}{\sigma}} \text{ for } p \to p_c,$$

where $\sigma = \frac{1}{2}$ and $2 < \tau \leq 3$.

Actually, we can determine τ as it is not independent of the critical exponents γ and σ determined previously. Let us evaluate the average cluster size S(p) assuming Eq. $n_s(p) \propto s^{-\tau} \exp\left(-\frac{s}{s_\xi}\right)$, for $s \gg 1$ is valid for all s. Neglecting the denominator which for $p \to p_c$ will approach p_c and thus be a constant we find,

$$\begin{aligned} S(p) &\propto \sum_{s=1}^{\infty} s^2 n_s(p) \\ &\propto \sum_{s=1}^{\infty} s^{2-\tau} \exp\left(-\frac{s}{s_\xi}\right) \\ &\approx \int_1^{\infty} s^{2-\tau} \exp\left(-\frac{s}{s_\xi}\right) ds \\ &= \int_{1/s_\xi}^{\infty} (zs_\xi)^{2-\tau} \exp(-z) s_\xi dz \quad \text{with } z = s/s_\xi \end{aligned}$$

$$= s_\xi^{3-\tau}\int_{1/s_\xi}^{\infty} z^{2-\tau}\exp(-z)dz$$
$$= s_\xi^{3-\tau}\int_{0}^{\infty} z^{2-\tau}\exp(-z)dz \quad \text{for } p \to p_c \text{ where } s_\xi \to \infty$$
$$= s_\xi^{3-\tau}\Gamma(3-\tau)$$

where Γ is the Gamma function,

$$\Gamma(x) = \int_0^\infty z^{x-1}\exp(-z)dz.$$

Example: Show that the Gamma function Γ defined by the integral above satisfies the recursion relation $\Gamma(x+1) = x\Gamma(x)$ and deduce that $\Gamma(x+1) = x!$ for x = 0, 1, 2,... where 0! = 1.

Using $s_\xi \propto (p-p_c)^{-2}$, we see that,

$$S(p) \propto (p-pc)^{2\tau-6} \propto (p-pc)^{-1},$$

where we have used Eq. $S(p) = 1+3T = \frac{1+p}{(1-2p)} = \frac{1+p}{2\left(\frac{1}{2}-p\right)} = \frac{1+p}{2(p_c-p)} = \Gamma_2(p_c-p)^{-1}$. Thus we must require that,

$$2\tau - 6 = -1 \Leftrightarrow \tau = \frac{5}{2}.$$

For the Bethe lattice,

$$n_s(p) \propto s^{-\frac{5}{2}}\exp\left(-\frac{s}{s_\xi}\right) \quad s \gg 1$$
$$s_\xi \propto |p-p_c|^{-2} \quad \text{for } p \to p_c.$$

The above procedure is a general way of deriving a relationship between critical exponents.

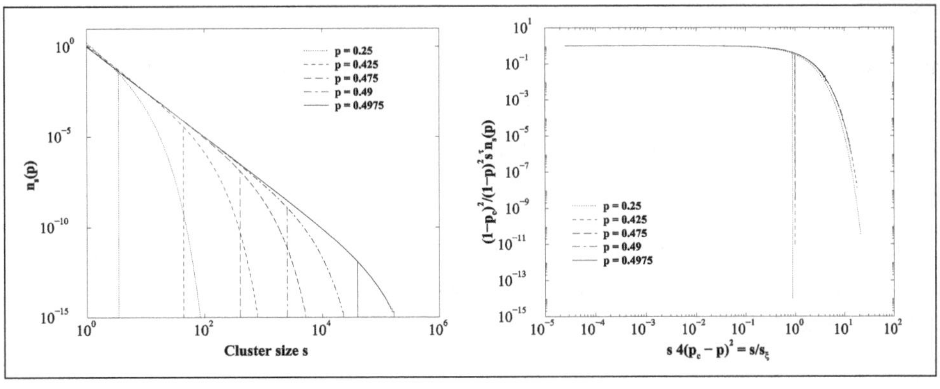

a) The cluster number distribution $n_s(p) = \left[\frac{(1-p)}{(1-p_c)}\right]^2 s^{-\tau}\exp\left(-\frac{s}{s_\xi}\right)$, equation

$\frac{n_s(p)}{n_s(p_c)} = \left[\frac{(1-p)}{(1-p_c)}\right]^2 \exp\left(-\frac{s}{s_\xi}\right)$ for different values of p approaching $p_c = 1/(z-1) = 1/2$. The vertical lines indicate the cutoff cluster size $s_\xi(p)$ for the various p values. The cluster number distribution is well approximated by a power law for $s \ll s_\xi$ while is decays rapidly for $s \gg s_\xi$. (b) By plotting $\left[\frac{(1-p_c)}{(1-p)}\right]^2 s^\tau n_s(p)$ versus $s/s_\xi \approx s \cdot 4(p_c - p)^2$, all the data collapses onto the scaling function e^{-x}.

τ

Now, we postulate a scaling ansatz (form) valid for all p and large s.

Ansatz 1

$$n_s(p) \propto s^{-\tau} \exp\left(-\frac{s}{s_\xi}\right), \quad s \gg 1,$$

$$s_\xi \propto |p-p_c|^{-\frac{1}{\sigma}} \quad p \to p_c,$$

or equivalently,

$$n_s(p) \propto s^{-\tau} \exp\left(-s|p-p_c|^{\frac{1}{\sigma}}\right) \quad s \gg 1 \text{ and } p \to p_c.$$

Note the role of s_ξ as a cutoff cluster size, where $n_s(p)$ is characterised by,

$$n_s(p) \propto \begin{cases} s^{-\tau} & \text{for } s \ll s_\xi \\ \text{decays rapidly} & \text{for } s \gg s_\xi, \end{cases}$$

representing the crossover from a behaviour of "critical clusters" (power-law distributed) to that of non-critical clusters. Note also that $n_s(p_c) = s^{-\tau}$ as $s_\xi = \infty$ at $p = p_c$.

We can, however, immediately identify two problems (a third problem will surface shortly) with the scaling ansatz 1.

(1) The scaling ansatz 1 fails to describe the 1d case. We found previously that,

$$n_s(p) = (1-p)^2 \exp\left(-\frac{s}{s_\xi}\right) = (p_c - p)^2 \exp\left(-\frac{s}{s_\xi}\right),$$

which is not a special case of our ansatz 1 since, instead of a power of s, we have $(p_c - p)^2$ in front of the exponential. You might be very tempted to say that $\tau = 0$, but that does actually not solve the problem, as we shall see below.

(2) For fixed s:

$$n_s(p) = \sum_{t=1}^{\infty} g_{s,t} p^s (1-p)^t$$

is a polynomial in p with a finite number of terms. Thus all derivatives of $n_s(p)$ with respect to p will remain finite for all p. However, using the scaling ansatz 1 for $p > p_c$ $n_s(p) = As^{-\tau}\exp\left(-s(p-p_c)^{\frac{1}{\sigma}}\right)$ we find, keeping s constant,

$$\frac{dn_s(p)}{dp} = As^{-\tau}\exp\left(-s(p-p_c)^{\frac{1}{\sigma}}\right)\frac{1}{\sigma}(p-p_c)^{\frac{1}{\sigma}-1}$$

and by differentiating once more,

$$\frac{d^2n_s(p)}{dp^2} = As^{-\tau}\left[\exp\left(-s(p-p_c)^{\frac{1}{\sigma}}\right)\left(\frac{1}{\sigma}(p-p_c)^{\frac{1}{\sigma}-1}\right)^2 + \exp\left(-s(p-p_c)^{\frac{1}{\sigma}}\right)\frac{1}{\sigma}\left(\frac{1}{\sigma}-1\right)(p-p_c)^{\frac{1}{\sigma}-2}\right].$$

In 2d, e.g., $\sigma = 0.4 \Leftrightarrow \frac{1}{\sigma} - 2 = 0.5$, so both the first and second derivative of $n_s(p)$ w.r.t. p remain finite as $p \to p_c$. The third derivative of $n_s(p)$, however, will contain a term with the factor $(p-p_c)^{\frac{1}{\sigma}-3}$ which will diverge as $p \to p_c$ contrary to the derivatives of Equation $n_s(p) = \sum_{t=1}^{\infty} g_{s,t}p^s(1-p)^t$.
We shall later resolve these two problems, so there is no need for you to worry too much at the moment.

Let us return to the question of deriving scaling relations assuming the scaling ansatz above for the cluster number distribution. Following the previous calculation of the average cluster size S(p) and noting that the Gamma function is just a number, we find,

$$S(p) \propto s_\xi^{3-\tau} \propto |p-p_c|^{\frac{\tau-3}{\sigma}}, \text{ for } p \to p_c,$$

as,

$$s_\xi \propto |p-p_c|^{-\frac{1}{\sigma}} \text{ for } p \to p_c.$$

However, by definition,

$$S(p) \propto |p-p_c|^{-\gamma} \text{ for } p \to p_c,$$

implying the scaling relation,

$$\gamma = \frac{3-\tau}{\sigma}.$$

We notice again that to ensure the divergence of S(p) for $p \to p_c$ we must have $\tau <$ in order to leave $\gamma > 0$. Also, we recover problem mentioned above because in 1d percolation theory $\gamma = 1$ and $\sigma = 1$ implying $\tau = 2$. However, there is no $s^{-\tau}$ term in our expression for the 1d cluster number distribution $n_s(p)$ and obviously the suggested "solution" of assuming $\tau = 0$ is not valid either.

Now we focus our attention on the percolation strength P(p), starting from equation. $P(p)+\sum_{s=1}^{\infty} sn_s(p)=p \quad \forall p$:

$$P(p)=p-\sum_{s=1}^{\infty} sn_s(p)$$

$$=p_c-\sum_{s=1}^{\infty} sn_s(p)+(p-p_c)$$

$$=\sum_{s=1}^{\infty} sn_s(p_c)-\sum_{s=1}^{\infty} sn_s(p)+\mathcal{O}(p-p_c)(\mathcal{O}(p-p_c)=\text{term of order}(p-p_c))$$

$$=\sum_{s=1}^{\infty} s^{1-\tau}-\sum_{s=1}^{\infty} s^{1-\tau}\exp\left(-\frac{s}{s_\xi}\right)+\mathcal{O}(p-p_c) \text{ using the scaling ansatz 1}$$

$$\propto \sum_{s=1}^{\infty} s^{1-\tau}\left\{1-\exp\left(-\frac{s}{s_\xi}\right)\right\}+\mathcal{O}(p-p_c).$$

$$\propto \sum_{s=1}^{\infty} s^{1-\tau}\left\{1-\exp\left(-\frac{s}{s_\xi}\right)\right\}\Delta s,$$

where we have dropped the term of order $(p-p_c)$ and introduced $\Delta s = 1$. When $p \to p_c$, the cutoff cluster size $s_\xi \to \infty$. Thus the term $\left\{1-\exp\left(-\frac{s}{s_\xi}\right)\right\} \approx 0$ for small s and the main contribution to the integral will come from large s of the order of s_ξ. We need to replace the summation by an integral. Going from a discrete sum to a continuous sum does not, of course, create a mathematical identity, unless one takes certain limits. However, when the main contribution from the sum is for large s, it is legitimate to approximate the sum by an integral $\sum \Delta s \approx \int ds$, as we are only interested in the the general scaling behaviour of the system. Thus,

$$P(p) \propto \int_1^{\infty} s^{1-\tau}\left\{1-\exp\left(-\frac{s}{s_\xi}\right)\right\} ds.$$

One might proceed by integration by parts. You will then discover that only one of the terms are non-zero. However, we will take another approach. We differentiate equation above w.r.t. $1/s_\xi$ to find,

$$\frac{dP(p)}{d\left(\frac{1}{s_\xi}\right)} \propto \int_1^{\infty} s^{2-\tau}\exp\left(-\frac{s}{s_\xi}\right) ds$$

$$\propto s_\xi^{3-\tau}\int_{1/s_\xi}^{\infty} z^{2-\tau}\exp(-z)\,dz \text{ substituting } z = s/s_\xi.$$

As we are ultimately interested in the limit $p \to p_c$ where the cutoff cluster size diverges, we replace the lower limit $1/s_\xi$ with zero and thus,

$$P(p) \propto \left(\frac{1}{s_\xi}\right)^{\tau-3}$$

so integrating we find,

$$P(p) = c_1 + c_2\left(\frac{1}{s_\xi}\right)^{\tau-2}$$

At $p = p_c$, $s_\xi = \infty$. But, $P(p_c) = 0$ implies that $c_1 = 0$. So,

$$P(p) \propto c_2\left(\frac{1}{s_\xi}\right)^{\tau-2},$$

$$\propto |p - p_c|^{\frac{\tau-2}{\sigma}}.$$

Hence, we have derived a new scaling relation,

$$\beta = \frac{\tau - 2}{\sigma}.$$

By definition,

$$P(p) = \begin{cases} 0 & p \le p_c \\ (p - p_c)^\beta & p \to p_c^+. \end{cases}$$

Now we find a third problem with our ansatz 1.

In the derivation above we did not assume $p > p_c$. Thus the ansatz 1 also predicts an infinite cluster for $p < p_c$, which is of course not correct.

Reflecting upon the three problems, we realise that they can all be resolved, as we shall see shortly, by avoiding the argument $s|p - p_c|^{\frac{1}{\sigma}}$ in the scaling ansatz. Thus let us assume the following general scaling form:

Ansatz 2

$$n_s(p) = q_0 s^{-\tau} f\left[q_1 (p - p_c) s^\sigma\right] \quad s \gg 1 \text{ and } p \to p_c.$$

The scaling ansatz 2 has the following properties:

- q_0 and q_1 are proportionality factors and pc the critical occupation probability. They all depend on the lattice details. They are non-universal.
- τ, σ are the critical scaling indices which are independent of p, p_c and the lattice structure. They are universal.

- f, the scaling function, is independent of p, p_c and detailed lattice structure. It is universal. The precise form of the scaling function f = f(z) has to be determined by computer simulations or other numerical methods since the function f is not predicted by the scaling assumption. Typically, however:

$$f(z)=\begin{cases}\text{constant} & |z|\ll 1 \Leftrightarrow |p-p_c|s^{\sigma}\ll 1 \Leftrightarrow s\ll |p-p_c|^{-\frac{1}{\sigma}} \propto s_{\xi} \\ \text{decays fast} & |z|\gg 1 \Leftrightarrow s\gg s_{\xi}.\end{cases}$$

that is, the scaling function f(z) reaches a constant value (except in 1d percolation theory) for $|z|\ll 1 \Leftrightarrow s\ll s_{\xi}$, and decays rather fast for $|z|\gg 1 \Leftrightarrow s\gg s_{\xi}$. Thus, the role of s_{ξ} as a cutoff and crossover size is maintained. Later, we shall look more carefully at the crossover properties.

- τ, σ and f depend on dimensionality.
- By plotting $\frac{1}{q_0}S^{\tau}n_s(p)$ versus $(p-p_c)q_1 s^{\sigma}$, all the data will fall on the universal curve outlining the scaling function f. This phenomenon is known as data collapse.

Problems above highlights some of the points above.

Let us check whether the general scaling ansatz 2 really resolves the three problems associated with the former scaling ansatz 1.

(1) The 1d cluster number distribution:

$$\begin{aligned} n_s(p) &= (p_c-p)^2\exp\left(-\frac{s}{s_{\xi}}\right) \\ &\approx (p_c-p)^2\exp(-s(p_c-p)) \quad \text{for } p\to p_c \\ &= s^{-2}(s(p_c-p))^2\exp(-s(p_c-p)) \\ &= s^{-2}f\left[s^{\sigma}(p_c-p)\right] \end{aligned}$$

which is of the form of the general scaling ansatz 2 with $\tau = 2$ and $\sigma = 1$ and f(x) = $x^2\exp(-x)$. Thus the scaling relation Equation $\gamma=\frac{3-\tau}{\sigma}$. is indeed fulfilled. Note, however, that unlike in higher dimensions, the scaling function increases like x^2 for $x \ll 1$ but that it indeed decays rapidly for $x \gg 1$.

(2) Keeping s constant we find,

$$\frac{\partial^k n_s(p)}{\partial p^k} = q_0 s^{-\tau} f^{(k)}\left[q_1(p-p_c)s^{\sigma}\right]q_1^k s^{\sigma k}$$

which is not divergent for $p\to p_c$ assuming that f is analytic, that is, all derivatives of f(z) with respect to z are finite everywhere, also at z = 0.

(3) Let us calculate the strength of the infinite cluster.

$$-P(p)=\sum_{s=1}^{\infty} sn_s(p)-p$$

$$=\sum_{s=1}^{\infty} sn_s(p)-p_c+(p_c-p)$$

$$=\sum_{s=1}^{\infty}\left[sn_s(p)-sn_s(p_c)\right]+\mathcal{O}(p-p_c)\ \text{ using}\sum sn_s(p_c)=p_c$$

$$\approx\int_1^{\infty} q_0 s^{1-\tau}\left(f\left[q_1(p-p_c)s^{\sigma}\right]-f[0]\right)ds$$

Now substitute $z=q_1(p-p_c)s^{\sigma}$. Notice that if $p > p_c$ the argument z is positive while for $p < p_c$ the argument is negative. The cluster size s, however, is always positive so $s=|z|^{\frac{1}{\sigma}}(q_1|p-p_c|)^{-\frac{1}{\sigma}}$ implying that ds $=\frac{1}{\sigma}|z|^{\frac{1}{\sigma}-1}(q_1|p-p_c|)^{-\frac{1}{\sigma}}$ dz with dz positive. Thus we find,

$$-P(p)=\int q_0|z|^{\frac{1-\tau}{\sigma}}(q_1|p-p_c|)^{\frac{\tau-1}{\sigma}}\left[f(z)-f(0)\right]\frac{1}{\sigma}|z|^{\frac{1}{\sigma}-1}(q_1|p-p_c|)^{-\frac{1}{\sigma}}dz$$

$$=q_0(q_1|p-p_c|)^{\frac{\tau-2}{\sigma}}\int|z|\frac{2-\tau}{\sigma}-1\left[f(z)-f(0)\right]\frac{1}{\sigma}dz.$$

Now it is important to realise that the limits of the integral will depend on the value of p since $z=q_1(p-p_c)s^{\sigma}$. If $p > p_c$, the upper limit will be ∞ (lower limit $p-p_c$ approaches 0 as $p\to p_c^+$) while for p < pc, the lower limit will be $-\infty$ (upper limit approaching 0 as $p\to p_c^-$).

Thus if,

$$\int|z|^{\frac{2-\tau}{\sigma}-1}\left[f(z)-f(0)\right]\frac{1}{\sigma}dz=\begin{cases}\text{constant}\neq 0 & \text{for } p>p_c\\ 0 & \text{for } p<p_c\end{cases}$$

we have solved the third problem as well, implying,

$$P(p)\propto\begin{cases}(p-p_c)^{\beta} & \text{for } p\to p_c^+\\ 0 & \text{for } p<p_c\end{cases}$$

$$\text{with }\ \beta=\frac{\tau-2}{\sigma}.$$

Discuss the qualitative behaviour of the scaling function f(z) as a function of $z = q_1(p - p_c)s^{\sigma}$ in order to satisfy the constraint given in equation.

$$\int|z|^{\frac{2-\tau}{\sigma}-1}\left[f(z)-f(0)\right]\frac{1}{\sigma}dz=\begin{cases}\text{constant}\neq 0 & \text{for } p>p_c\\ 0 & \text{for } p<p_c\end{cases}.$$

Cluster Structure

Till now we have only considered the distribution of cluster sizes and have come up with some useful scaling laws and scaling relations. Now we turn our attention to the geometry of the clusters. We shall do so by mainly studying the fractal geometry, which contains information about the density of the clusters on different length scales.

Cluster Radius and Fractal Dimension

Let r_i denote the position of the ith occupied site in a cluster of size s.

Definition: The centre of mass of an s-cluster $r_{cm} = \frac{1}{s}\sum_{i=1}^{s} r_i$.

Definition: The radius of gyration R_s of a given s-cluster is defined by,

$$R_s^2 = \text{average square distance to the centre of mass}$$

$$= \left\langle |r_i - r_{cm}|^2 \right\rangle$$

$$= \frac{1}{s}\sum_{i=1}^{s} |r_i - r_{cm}|^2$$

We want to prove that, $R_s^2 = \frac{1}{2}$ average square distance between two cluster sites $= \frac{1}{2}\frac{1}{s^2}\sum_{ij} |r_i - r_j|^2$.

By definition, the lefthand side of equation above,

$$R_s^2 = \frac{1}{s}\sum_{i=1}^{s} |r_i - r_{cm}|^2$$

$$= \frac{1}{s}\sum_{i=1}^{s} (r_i \cdot r_i + r_{cm} \cdot r_{cm} - 2 r_i \cdot r_{cm})$$

$$= \frac{1}{s}\sum_{i=1}^{s} r_i \cdot r_i + \frac{1}{s}\sum_{i=1}^{s} r_{cm} \cdot r_{cm} - \frac{2}{s}\sum_{i=1}^{s} r_i \cdot r_{cm}$$

$$= \frac{1}{s}\sum_{i=1}^{s} r_i \cdot r_i + r_{cm} \cdot r_{cm} - 2\left(\frac{1}{s}\sum_{i=1}^{s} r_i\right) \cdot r_{cm}$$

$$= \frac{1}{s}\sum_{i=1}^{s} r_i \cdot r_i - r_{cm} \cdot r_{cm}.$$

Let us rewrite the righthand side of Equation above,

$$R_s^2 = \frac{1}{s}\sum_{i=1}^{s} |r_i - r_{cm}|^2$$

$$= \frac{1}{2}\frac{1}{s^2}\sum_{i=1}^{s}\sum_{j=1}^{s} \left(r_i \cdot r_i + r_j \cdot r_j - 2 r_i \cdot r_j\right)$$

$$= \frac{1}{2}\frac{1}{s^2}\left[\sum_{i=1}^{s} s\, r_i \cdot r_i + \sum_{j=1}^{s} s\, r_j \cdot r_j - 2\left(\sum_{i=1}^{s} r_i\right) \cdot \left(\sum_{j=1}^{s} r_j\right)\right]$$

$$= \frac{1}{s}\sum_{i=1}^{s} r_i \cdot r_i - \frac{1}{s^2}(s\, r_{cm}) \cdot (s\, r_{cm})$$

$$= \frac{1}{s}\sum_{i=1}^{s} \mathbf{r}_i \cdot \mathbf{r}_i - \mathbf{r}_{cm} \cdot \mathbf{r}_{cm}.$$

The radius of gyration is a more useful quantity to work with since in many situations (like in Polymer Science) we have to deal with more complicated structures than straight lines, squares or spheres.

We also remember that the correlation function g(r) is defined as the probability that a site at a position r from an occupied site belongs to the same finite cluster. We have the identity:

$$\sum_{r} g(r) = S(p)$$

because the sum is the average number of sites to which an occupied site is connected.

Definition: The correlation length ξ is defined as:

$$\xi^2 = \frac{\sum_r r^2 g(r)}{\sum_r g(r)}.$$

The correlation length represents some average distance of two sites belonging to the same cluster.

For a given s cluster, $2R_s^2$ is the average squared distance between two cluster sites. The probability of a site being part of an s cluster is $sn_s(p)$, and it will be connected to s sites (if we include self-connection). The corresponding average over $2R_s^2$ gives the squared correlation length,

$$\xi^2 = \frac{\sum_s 2R_s^2 s^2 n_s(p)}{\sum_s s^2 n_s(p)}.$$

The correlation length ξ is the upper cutoff of the radius of those clusters which contribute to the mean cluster size near the percolation threshold. It is expected that ξ diverges as $p \to p_c$, like,

$$\xi \propto |p - p_c|^{-\nu}.$$

We have introduced a new critical exponent, ν, and we wish to determine how it is related to τ and σ. For 2d percolation, plausible but not rigorous arguments give $\nu = \frac{4}{3}$, in excellent agreement with numerical results. In 3d, ν is somewhat smaller than 0.9, whilst for Bethe lattices one has $\nu = \frac{4}{2}$, analogous to the mean-field theories for thermal phase transitions.

As we have already seen, many quantities diverge at the percolation threshold. Most of these involve sums over all cluster sizes s; their main contribution comes from s of the order $s_\xi \propto |p - p_c|^{-\frac{1}{\sigma}}$. Now we see that the correlation length ξ, which is also one of these quantities, is simply the radius of those clusters which contribute the most to divergence. This effect is the foundation of scaling theory. There is one and only one length ξ dominating the critical behaviour in an infinite lattice.

Finite Boxing of Percolating Clusters

We now want to find out how the R_s varies with s at the percolation threshold $p = p_c$ where $\xi = \infty$. Let M(L) denote the mass of the percolating cluster within linear distance L. If the percolating cluster was a compact object, then it would be of the form $M(L) \propto L^2$. However, at $p = p_c$, the percolating cluster is a fractal object. Figure below displays the size of the largest cluster S_∞.

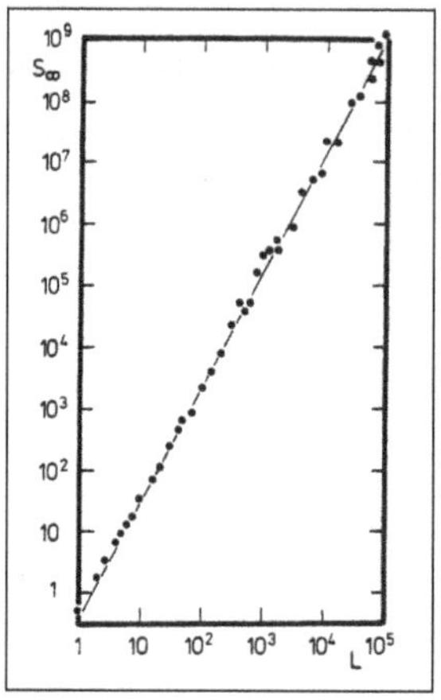

The size of the largest cluster S_∞ at $p = p_c$ as a function of the lattice size L.

(the percolating cluster) as a function of lattice size L in a double-logarithmic plot. The data are consistent with the existence of a fractal dimension D such that,

$$M(L) \propto L^D.$$

For the 2d case $D = \frac{91}{48} < 2$ and in 1d, D = 1.

Having obtained the relationship in ($M(L) \propto L^D$) it is natural to assume,

$$s \propto R_s^D \quad \text{for } s \gg a \text{ at } p = p_c.$$

Thus,

$$R_s \propto s^{\frac{1}{D}} \text{ for } p \to p_c.$$

Inserting Equation above into Equation $\xi^2 = \frac{\sum_s 2R_s^2 s^2 n_s(p)}{\sum_s s^2 n_s(p)}$ we have,

$$\xi^2 \propto \frac{\sum_s 2s^2 + \frac{2}{D} n_s(p)}{\sum_s s^2 n_s(p)}$$

$$\propto \frac{|p-p_c|^{\tau-3-\frac{(2)}{D\sigma}}}{|p-p_c|^{\tau-1-\frac{2}{\sigma}}}$$

$$\propto |p-p_c|^{-\frac{2}{D\sigma}}$$

$$\propto |p-p_c|^{-2\nu},$$

where we have used,

$$\sum_{s=1}^{\infty} s^k n_s(p) \propto |p-p_c|^{\frac{\tau-1-k}{\sigma}}.$$

Therefore we end up with the scaling relation,

$$\nu = \frac{1}{D\sigma} \Leftrightarrow D = \frac{1}{\nu\sigma}.$$

The behaviour at $p = p_c$ and the behaviour for $s \gg s_\xi$, or $R_s \ll \xi$ are indistinguishable. Therefore, ξ is the crossover length between a critical and noncritical behaviour. On length scales much less than ξ we have $n_s \propto s^{-\tau}$ and $R_s \propto s^{\frac{1}{D}}$ but for length scales much larger than ξ the scaleless behaviour disappears. Also,

$$s_\xi \propto \xi^D \propto |p-p_c|^{-\nu D} \propto |p-p_c|^{-\frac{1}{\sigma}}$$

confirming once again the scaling relation above between D, ν and σ.

If, however, $p \gg p_c$, the fractal dimension D is expected to be equal to the Eucledian dimension d of the lattice. So,

$$D = d \quad \text{for} \quad p > p_c.$$

Clusters above p_c are not fractals but 'normal' objects with D = d, provided $s \gg s_\xi$ or equivalently $R_s \gg \xi$. Thus s_ξ or ξ sets the cluster size and linear scale where there is a crossover from fractal to non-fractal behaviour.

Table: The critical exponents for the percolation theory problems in dimensions d = 1, 2, 3, 4, 5, 6 and in the Bethe lattice.

Exponent	1d	2d	3d	4d	5d	6d	Bethe
α	1	−2/3	-0.62	-0.72	-0.86	-1	-1
β	0	5/36	0.41	0.64	0.84	1	1
γ	1	43/18	1.80	1.44	1.18	1	1
ν	1	4/3	0.88	0.68	0.57	1/2	1/2
σ	1	36/19	0.45	0.48	0.49	1/2	1/2
τ	2	187/91	2.18	2.31	2.41	5/2	5/2
$D(p = p_c)$	1	91/48	2.53	3.06	3.54	4	4

Mass of the Percolating Cluster

Now we wish to see how the scaling relation of the mass M(L) changes for the two different cases.

(1) For $L \ll \xi$: (always the case at $p = p_c$, where $\xi = \infty$) The percolating cluster appears fractal, implying $M(L) \propto L^D$.

(2) For $L \ll \xi$: The cluster appear homogeneous.

$$M(L) = (\text{No. of lattice sites}) \times (\text{Prob. site belongs to percolating cluster})$$

$$= L^d P(p)$$

$$= L^d (p - p_c)^\beta$$

$$= L^d \xi - \frac{\beta}{\nu} \text{ since } \xi \propto |p - p_c|^{-\nu}.$$

Now consider $L \approx \xi$, that is, we match the observations above by substituting L with ξ:

$$M(L) \propto L^D \propto \xi^D,$$

and,

$$M(L) \propto L^D \xi^{-\frac{\beta}{\nu}} \propto \xi^{d-\frac{\beta}{\nu}}.$$

Therefore, we can state that,

$$D = d - \frac{\beta}{\nu},$$

which is know as a hyperscaling relation because the Euclidean dimension d enters in the scaling relation.

Since the percolating cluster has a constant density for $L \gg \xi$, it is natural to divide the system into boxes of linear size ξ. In d dimensions, the total volume L^d will be divided into $\left(\frac{L}{\xi}\right)^d$ boxes. Since the cluster inside each of these boxes of size ξ^d has a mass of order ξ^D, the total mass of the cluster is given by,

$$M(L,\xi) = \left(\frac{L}{\xi}\right)^d \xi^{D-d} L^d,$$

which is, of course, equivalent to $P(p)L^d$.

We have,

$$M(L,\xi) \propto \begin{cases} L^D & L \gg \xi \\ \xi^D \left(\frac{L}{\xi}\right)^d & L \gg \xi \end{cases}$$

which can be written in an alternative form of,

$$M(L,\xi) = L^D m\left(\frac{L}{\xi}\right),$$

where the scaling function,

$$m(x)\begin{cases} \text{constant} & x = \dfrac{L}{\xi} \ll 1 \\ x^{d-D} & x = \dfrac{L}{\xi} \gg 1. \end{cases}$$

The argument x of the scaling function is a dimensionless number, namely the ratio of the two length scales $\frac{L}{\xi}$. No other length scales play a role. The scaling function describes a crossover from fractal behaviour at length scales L much smaller than the correlation length ξ to uniform behaviour at length scales L much larger than the correlation length ξ. In Nature, there are many examples of crossover phenomena from one type of behaviour to another. Take as an example the table in front of you. On large length scales, the surface of the table is smooth, but going to very small length scales, the surface will become very rough and probably fractal. The mountain range is yet another example. Far away, i.e., on large length scales, it looks pretty smooth, but, as you are well aware, if you are in the middle of the mountain range, on small length scales, it is pretty rough and rugged.

Example: Give one or two more examples of such crossover phenomena in nature.

The density has the form,

$$\rho(L,\xi) = \frac{M(L,\xi)}{L^d} \propto \begin{cases} L^{D-d} & L \ll \xi \\ \xi^{D-d} & L \gg \xi \end{cases} = L^{D-d}\,\tilde{\rho}\left(\frac{L}{\xi}\right),$$

with a scaling function,

$$\tilde{\rho}(x) = \begin{cases} \text{constant} & x - \dfrac{L}{\xi} \ll 1 \\ x^{d-D} & x - \dfrac{L}{\xi} \gg 1. \end{cases}$$

To be aware that the density actually decreases with the length scale when $L < \xi$ can be of immense practical importance. Imagine that you have just been hired by an oil company and you were given a sample of dimensions 0.1 × 0.1 × 0.1 m^3 of the porous medium where the oil resides and had to estimate how much oil the company could expect to recover in the oil field of dimension, say, 100 × 100 × 100 km^3. The percolation model is used as a model for the distribution of oil and gas inside porous rock in oil reservoirs. The empty sites model hard rock while the occupied sites are pores filled with oil. The average concentration of oil is p. Now, if $p < p_c$ there are only finite clusters, so let us assume $p > p_c$. The strength P(p) will be the probability of drilling into a percolating cluster, but how much oil can you recover? You would head for an immediate dismissal if you argued as follows: Measure the density of oil in the sample $\rho(L_1 = 0.1m)$. The mass of oil to be recovered is $M = \rho(L_1 = 0.1m) \cdot (100000m)^3$. The argument is wrong simply because the density of oil at length scales $L_2 = 100$ km is not given by the density of oil at length scales $L_1 = 0.1$ m, as shown in figure below. The proper way to argue would be the following: measure the density of oil in the sample $\rho(L_1 = 0.1m) = CL_1^{D-d}$. The density of oil at the reservoir length scale $\rho(L_2 = 100\,km) = CL_2^{D-d}$ with the same constant C. Taking the ratio we find,

$$\begin{aligned}\rho(L_2 = 100\,km) &= \left(\frac{L_2}{L_1}\right)^{D-d} \rho(L_1 = 0.1m)\\ &= \left(10^6\right)^{D-d} \rho(L_1 = 0.1m)\\ &\approx 1.5 \cdot 10^{-3} \rho(L_1 = 0.1m),\end{aligned}$$

since in 3d $D - d \approx 2.53 - 3 = -0.47$. The former estimate is wrong by a factor 1000.

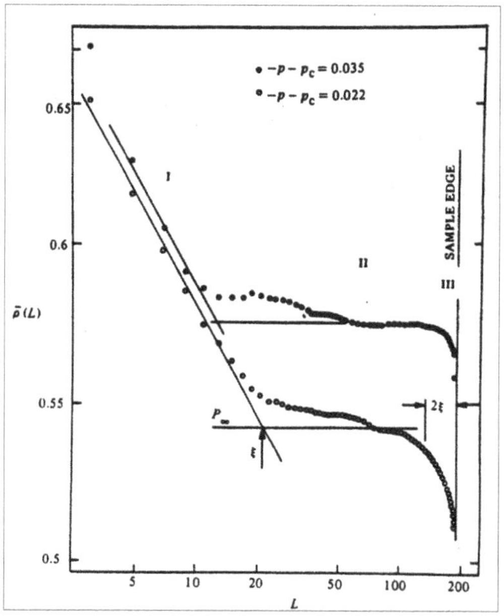

The density of sites belonging to the percolating cluster within a region of linear size L. In region I, the ratio $L/\xi \ll 1$ and in region II, $L/\xi \gg 1$. Region III is related to the fact, that the lattices used are finite.

Fractals

The percolating cluster at $p = p_c$ is an example of a random fractal. When $p \neq p_c$ large clusters appear fractal on length scales up to the correlation length $\xi \propto |p - p_c|^{-\nu}$, roughly speaking. In mathematics, we can also encounter deterministic fractals as for example the Sierpinski carpet and the Sierpinski gasket.

For a geometrical fractal, the mass scales with the linear size raised to the fractal dimension,

$$M(L) = L^D \Leftrightarrow D = \frac{\log M(L)}{\log L}.$$

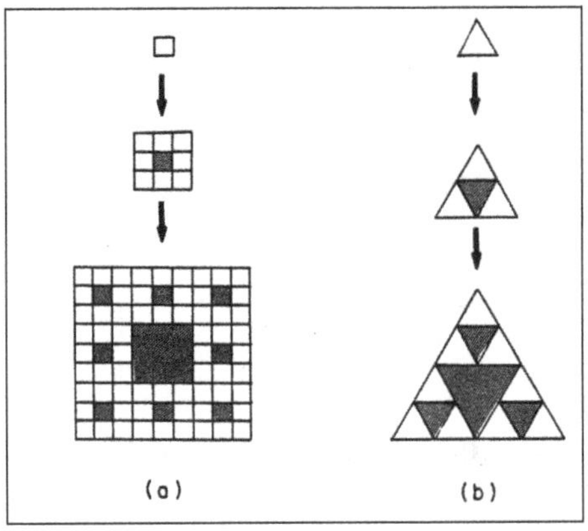

Figure shows two examples of deterministic fractals. (a) The Sierpinski carpet. The first three levels in the algorithm for constructing the Sierpinski carpet are shown. The linear scale is enlarged by a factor 3 with the condition that an occupied square (white) is replaced with with 3 × 3 squares in which the centre square is empty (grey) while an empty square (grey) is replaced with 3 × 3 empty squares. (b) The Sierpinski gasket. A similar algorithm generates the Sierpinski gasket.

Let n denote the iteration number. By inspection of the Sierpinski carpet, $M = 8^n$ and $L = 3^n$, implying,

$$D = \frac{\log 8^n}{\log 3^n} = \frac{\log 8}{\log 3} \approx 1.893.$$

In the case of the Sierpinski gasket, $M = 3^n$ and $L = 2^n$, yielding,

$$D = \frac{\log 3}{\log 2} \approx 1.585.$$

Note that if the length scale is rescaled by a factor b, then,

$$M\left(\frac{L}{b}\right) = \left(\frac{L}{b}\right)^D = \frac{L^D}{b^D} = \frac{M(L)}{b^D}$$

so that the mass is reduced by a factor b^D. Using this observation as the characteristic of fractal behaviour for random fractals we can easily find, say, the scaling of the mass of the percolating cluster $M(L, \xi)$ as a function of L and ξ. Let us rescale the length scales by a factor b, that is, $L \rightarrow \frac{L}{b}$ and $\xi \rightarrow \frac{\xi}{b}$ (no other length scales are present). Thus:

$$\begin{aligned} M(L,\xi) &= b^D M\left(\frac{L}{b}, \frac{\xi}{b}\right) \\ &= b^D b^D M\left(\frac{L}{b^2}, \frac{\xi}{b^2}\right) \\ &\vdots \\ &= b^{Dl} M\left(\frac{L}{b^l}, \frac{\xi}{b^l}\right). \end{aligned}$$

If we are at the critical point $p = p_c, \xi \propto |p - p_c|^{-\nu} = \infty$, but then $\frac{\xi}{b^l} = \infty \ \forall l$. If we stop the above renormalisation procedure when $b^l = L$ we find,

$$\begin{aligned} M(L, \xi = \infty) &= \left(b^l\right)^D M\left(\frac{L}{b^l}, \frac{\xi}{b^l} = \infty\right) \\ &= L^D M(1, \infty) \\ &= L^D. \end{aligned}$$

This proves that self-similarity is mathematically expressed by power-law behaviour. Now, assume $p > p_c$ with $\xi \propto |p - pc|^{-\nu} < \infty$ with,

$$M(L, \xi) = b^{Dl} M\left(\frac{L}{b^l}, \frac{\xi}{b^l}\right).$$

We now have to consider two cases depending on the ratio L/ξ.

(a) If $L/\xi \ll 1 \Leftrightarrow L \ll \xi$ (which is allways the case at p = pc) we stop the iteration process when we reach the smallest length scale $b^l = L$ and we "recover":

$$M(L, \xi) = L^D M\left(1, \frac{\xi}{L}\right) = L^D M(1, \infty) \propto L^D.$$

(b) If $L/\xi \gg 1 \Leftrightarrow L \gg \xi$, we stop the iteration process when $b^l = \xi$ and we find,

$$M(L, \xi) = \xi^D M\left(\frac{L}{\xi}, 1\right) \propto \xi^D \left(\frac{L}{\xi}\right)^d = \xi^{D-d} L^d,$$

since for a uniform system,

$$M\left(\frac{L}{\xi}, 1\right) \propto \left(\frac{L}{\xi}\right)^d.$$

Finite-size Scaling

Let us consider the percolation problem for p ≠ pc where the correlation length is finite,

$$\xi \propto |p - p_c|^{-\nu} \Rightarrow |p - p_c| \propto \xi^{-\frac{1}{\nu}}.$$

The strength of the infinite cluster for $p > p_c$ can then be expressed in terms of the finite correlation length,

$$P(p) \propto (p - p_c)^\beta \propto \xi^{-\frac{\beta}{\nu}}$$

and similarly for the average cluster size,

$$S(p) \propto |p - p_c|^{-\gamma} \propto \xi^{\frac{\gamma}{\nu}}.$$

These are the results in infinite lattices with $L=\infty$. Now, what happens at finite lattice sizes when $L<\infty$? Well, nothing happens as long as $L \gg \xi$. The only relevant length scale is the correlation length ξ. However, when $L \ll \xi$, the length scale will be set by L. This we have already encountered before, we just didn't stress that point. Better late that never:

$$P(L,\xi)=\rho(L,\xi)$$
$$=\frac{M(L,\xi)}{L^d}$$
$$=\begin{cases} L^{D-d} & L \ll \xi \\ \xi^{D-d} & L \gg \xi. \end{cases}$$
$$=\begin{cases} L^{-\frac{\beta}{\nu}} & L \ll \xi \\ \xi^{-\frac{\beta}{\nu}} & L \gg \xi. \end{cases}$$
$$=\xi^{-\frac{\beta}{\nu}} f\left(\frac{L}{\xi}\right)$$

where the scaling function,

$$f\left(\frac{L}{\xi}\right)=\begin{cases} \left(\frac{L}{\xi}\right)^{-\frac{\beta}{\nu}} & \frac{L}{\xi} \ll 1 \\ \text{constant} & \frac{L}{\xi} \gg 1. \end{cases}$$

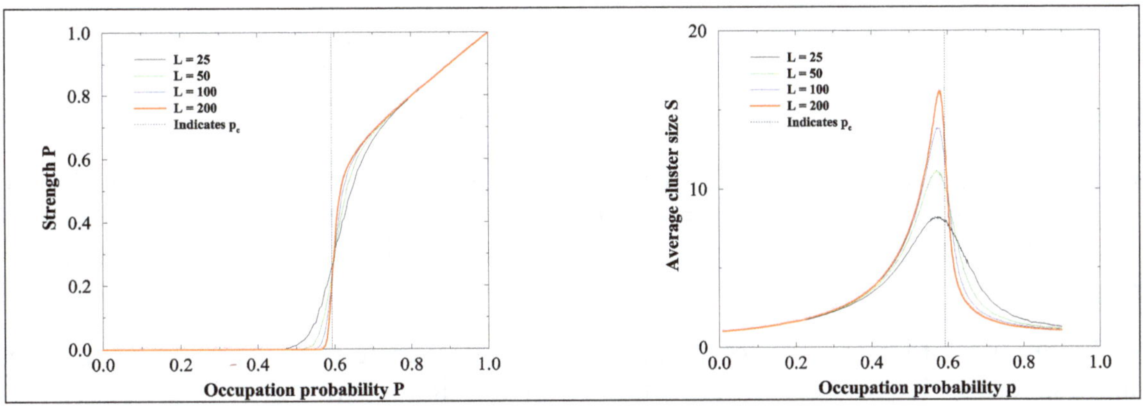

Figure above shows (a) The strength P(p) measured in a 2d square lattice for finite system sizes as a function of occupation probability p. When p is far away from $p_c, L \gg \xi$ and the strength is determined by $\xi^{-\frac{\beta}{\nu}}$. However, with p closer to pc, the finite-size effect is seen and P(p) decays with system size as $L^{-\frac{\beta}{\nu}}$. (b) The average cluster size S(p) measured in a 2d square lattice for finite system sizes L as a function of occupation probability p. When p is far away from $p_c, L \gg \xi$ and the average cluster size is determined by the correlation length $\xi^{\frac{\gamma}{\nu}}$. However, with p closer to p_c, the finite-size effect is seen and S(p) increases with system size as $L^{\frac{\gamma}{\nu}}$.

Similarly for the average cluster size, we would expect,

$$S(L,\xi)\propto\begin{cases}L^{\frac{\gamma}{\nu}} & L\ll\xi\\ \xi^{\frac{\gamma}{\nu}} & L\gg\xi.\end{cases}$$

Generally, if we have a quantity,

$$X\propto|p-p_c|^{-\chi}\propto\xi^{\frac{\chi}{\nu}}\ \text{for}\ L\gg\xi$$

then we would expect,

$$X(L,\xi)\propto\begin{cases}L^{\frac{\gamma}{\nu}} & L\ll\xi\\ \xi^{\frac{\gamma}{\nu}} & L\gg\xi.\end{cases}$$
$$=\xi^{\frac{\chi}{\nu}}X_1\left(\frac{L}{\xi}\right)$$

or, in terms of occupation probabilities:

$$X(L,p)=|p-p_c|^{-\chi}X_2\left(L^{\frac{1}{\nu}}(p-p_c)\right).$$

The important message we get is that by studying finite-size scaling, that is, studying X as a function of (finite) system sizes L at $p = p_c$, we can extract the exponent $\frac{\chi}{\nu}$.

Real Space Renormalisation in Percolation Theory

Imagine a percolation lattice with n sites. The total number of different microscopic configurations is 2^n as a site can be in one of two states – occupied with probability p or empty with probability 1 – p. Table shows the number of microscopic configurations in a two dimensional square lattice of linear size L.

Table: The number of microscopic configurations 2 n as a function the linear size L of a two dimensional square lattice where $n = L^2$, the number of sites in the lattice. For lattice size L = 17, the number exceeds the estimated number of atoms in the universe $\approx 10^{81}$.

L	1	2	3	4	5	6	7	8	9	10	17	100
n	1	4	9	16	25	36	49	64	81	100	289	10000
2^n	2	16	512	65536	$3\cdot10^7$	$7\cdot10^{10}$	$6\cdot10^{14}$	$2\cdot10^{19}$	$2\cdot10^{24}$	10^{30}	10^{87}	10^{3000}

In principle, if we wanted to calculate the average of a quantity $\langle A\rangle$ (say the average cluster size), we should evaluate the quantity Ai associated with each configuration i and then weight the quantity A_i with the probability pi (which would be of the form $p^m (1 - p)^{(n-m)}$, where m is the number of occupied sites in that particulate configuration) of the system to be in that microscopic state, that is,

$$\langle A \rangle = \frac{\sum_i p_i A_i}{\sum_i p_i} = \sum_i p_i A_i.$$

This procedure, however, soon becomes an impossible task due to the numbers of different states involved. With today's computing power one might do, say, a 7 × 7 system, but an 8 × 8 system would take approximately 40 years on a tera flop (10^{12} calculations per second) while the time required to do a 10 × 10 system would exceed the age of the universe which is approximately $15 \cdot 10^9$ years.

Real space renormalisation is a coarse graining procedure where we systematically reduce the number of degrees of freedom (sites) in order to break a large problem down into a sequence of smaller and more manageable stages by eliminating fluctuations on scales less than a given length scale b. The effect will be to reveal the large scale behaviour of the system.

Before we introduce the real space renormalisation method, let us make some observations related to the phase transition in a percolation model.

- The empty state at occupation probability p = 0 is self-similar with an associated correlation length $\xi = 0$.
- The fully occupied state at occupation probability p = 1 is self-similar. The associated correlation length $\xi = 0$. The ordered state has an order parameter P(p = 1) = 1.
- The states at the critical occupation probability $p = p_c$ are also self-similar and the correlation length $\xi = \infty$. The infinite cluster is fractal and looks alike on all length scales and we have clusters of all sizes from s = 1 to $s = \infty$.

Loosely speaking, the correlation length ξ sets the scale of the largest fluctuation from the "averaged" state. For $p \neq p_c$, the correlation length is identified with the linear size of the cutoff cluster size s_ξ.

When $p < p_c$ the upper cutoff of the linear size of the fluctuations away from the empty state is given by ξ and when p approaches zero, ξ approaches zero. At p = 0, there are no clusters left at all and $\xi = 0$.

When $p > p_c$, the correlation length can also be identified with the upper cutoff in the linear size of the largest holes in the percolating (spanning) cluster which indicates how far away you are from the fully occupied state. When p approaches one, ξ approaches zero. At p = 1, there are no holes left and $\xi = 0$.

At the phase transition point $p = p_c$, the correlation length is infinite. Thus there are fluctuations from the smallest length scale up to infinity.

The real space renormalisation technique is based on the so-called block site (spin) technique introduced by Leo Kadanoff and later formalised into the renormalisation group method by Kenneth

Wilson, who received the Nobel price in 1982 for "his theory for critical phenomena in connection with phase transitions". The renormalisation group method has three basic steps:

- Divide the lattice into blocks of linear size b (in terms of the lattice constant) with each block containing a few sites (spin).
- Next, the coarse graining procedure takes place. The sites in the blocks are averaged in some way (to be specified more precisely shortly) and the entire block is replaced by a single super site (spin) which is occupied with a probability according to the renormalisation group transformation $p' = R_b(p)$.

  In the combined procedure 1 and 2, one should keep the symmetry of the original lattice such that we can repeat the coarse graining procedure again. The result of these two operations are to create a new lattice whose fundamental spacing is b times as large as the original lattice.
- Restore original lattice constant by rescaling the length scales by the factor b.

These 3 steps define a renormalisation group transformation R_b, where the number of degrees of freedom is reduced drastically. Note that it is called a group because of the property that when applying the transformation to a configuration $\{s_i\}$,

$$R_{b2}\left(R_{b1}\left(\{s_i\}\right)\right)=R_{b1b_2}\left(\{s_i\}\right)$$

but it is of course not a group in a strictly mathematical sense because no inverse transformation exists since we are reducing the number of degree of freedom.

The effect of the coarse graining procedure in step 2 is to eliminate from the system all fluctuations whose scale is smaller than the block size b. Any small scale fluctuations of the sites over a range of less than b lattice units will be smeared out. This is somewhat similar to the process taking place if you view the system through a lens out of focus – all the smaller features are blurred but the large scale features are unaffected.

Let us clarify the 3 steps by looking at a simple example. Consider a 2d square lattice with 9×9 sites, that is, 81 degrees of freedom and a total of $2^{81} = 2 \cdot 10^{24}$ different configurations.

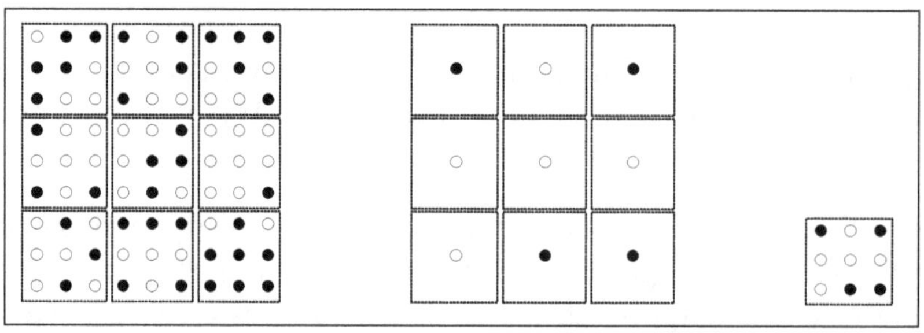

The three renormalisation group steps illustrated in a 2d square lattice with b = 3.
The majority rule defines the renormalisation group transformation.

- Divide the square lattice up into small 3×3 blocks with a total of 9 sites (b = 3.)
- We now have to specify the course graining procedure which will define the renormalisation group transformation. One possibility would be to use the so-called majority rule which states that if a majority of the sites in the block is occupied, then the super site should be

occupied, otherwise empty. The probability, that a majority of sites are occupied in the 3×3 block then defines the renormalisation group transformation,

$$p' = R_b(p) = p^9 + K_{9,1}p^8(1-p) + K_{9,2}p^7(1-p)^2 + K_{9,3}p^6(1-p)^3 + K_{9,4}p^5(1-p)^4,$$

where $K_{9,k}$ is the number of different ways to place k empty sites into the cells of 9 sites. The super site should be occupied with probability $p' = R_b(p)$ and empty with probability $1-p' = 1-R_b(p)$.

- Rescale the lattice of super sites by a factor 3 to restore original lattice constant. Let ξ denote the correlation length in the original lattice. Assume we are close to p_c. Then,

  $$\xi = \text{constant}|p-p_c|^{-v}$$

We denote the correlation length in the new renormalised lattice by ξ'. Since the length scales have been rescaled by a factor b,

$$\xi' = \frac{\xi}{b} \Leftrightarrow$$

$$\text{constant}|R_b(p)-p_c|^{-v} = \frac{\text{constant}|p-p_c|^{-v}}{b}$$

where $R_b(p)$ is the renormalisation group transformation that determines the new occupation probability in the rescaled lattice associated with the new (smaller as b > 1) correlation length.

Simple rearranging implies that,

$$\left(\frac{|R_b(p)-p_c|}{|p-p_c|}\right)^{-v} = \frac{1}{b}$$

so the critical exponent v is given by,

$$v = \frac{\log b}{\log\left(\frac{|R_b(p)|-p_c}{|p-p_c|}\right)}.$$

The basic idea is that at the critical point p_c we have self-similarity, that is,

$$\xi' = \xi.$$

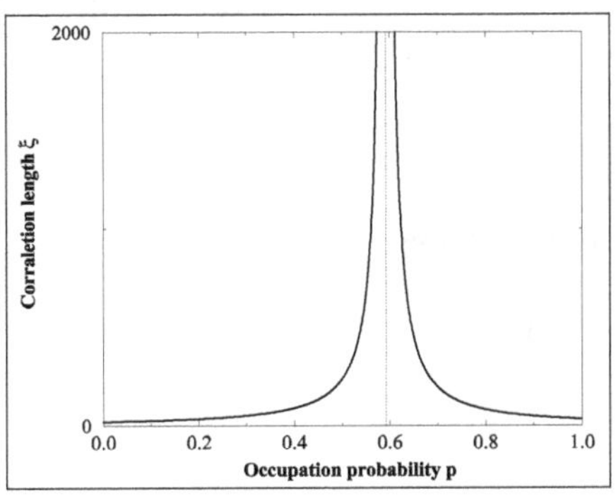

When renormalising the length scales by a factor b, the correlation length decreases unless it was 0 or ∞ initially. Associated with a decrease in the correlation length is a flow in occupation probability space either towards p = 0, if initially $p < p_c$ or towards p = 1, if initially $p > p_c$. The fixed points of the renormalisation group transformation are associated with the trivial self-similar states at p = 0 or p = 1 or the nontrivial self-similar states at $p = p_c$.

This equation can only be consistent with Equation $\left(\frac{|R_b(p)-p_c|}{|p-p_c|}\right)^{-v} = \frac{1}{b}$ if,

$$\xi = \begin{cases} \infty \\ 0. \end{cases}$$

This is intuitively clear as $\xi = 0$ is associated with either p = 0 (empty lattice) or p = 1 (fully occupied lattice) the two trivially self-similar cases and $\xi = \infty$ with the critical point pc. Furthermore, we can also conclude that,

$$\xi = \text{constant } |p-p_c|^{-v} = \xi' = \text{constant } |R_b(p)-p_c|^{-v}$$

which implies that the phase transition $\xi = \infty$ is identified with (one of) the fixed points p ? of the renormalisation group transformation, that is, the solutions of the equation,

$$Rb(p^*) = p^*.$$

Thus,

$$\frac{|R_b(p)-p_c|}{|p-p_c|} = \frac{|R_b(p)-R_b(p_c)|}{|p-p_c|}$$
$$= \frac{dR_b(pc)}{dp} \text{ for } p \to p_c$$

leaving, see Equation $v = \frac{\log b}{\log\left(\frac{|R_b(p)|-p_c}{|p-p_c|}\right)}$.

Renormalisation Group Transformation in 1d

Consider a renormalisation group transformation in 1d. Take a 1d lattice where each site is occupied with probability p. Divide the lattice into blocks with b sites. Let the renormalisation group transformation be determined by the probability of having a spanning cluster. The probability of having a spanning cluster in a block of b sites is,

$$R_b(p) = p^b.$$

The fixed point equation can easily be solved,

$$R_b(p^\star) = p^{\star b} = p^\star \Leftrightarrow p^\star = \begin{cases} 0 \\ 1. \end{cases}$$

If we start with an empty lattice p = 0, the renormalised lattice will also be empty, associated with the fixed point $p^\star = 0$. If we start with a fully occupied lattice p = 1, the renormalised lattice will also be fully occupied since all the blocks contain a percolating cluster, associated with the fixed point $p^\star = 1$. However, if we start with a lattice containing some empty sites p < 1, the renormalised lattice will contain even more empty sites because $p' = R_b(p) = p^b < p$. Repeating the renormalisation procedure will gradually bring the renormalised occupation probability towards zero. Clearly, $p^\star = 0$ is associated with $\xi = 0$, while $p^\star = 1$ is associated with $\xi = \infty$ and must thus be identified with the critical occupation probability p_c.

Furthermore, in order to calculate the correlation length exponent ν we need to take the derivative of the renormalisation group transformation evaluated at $p^\star = 1$.

$$\frac{dR_b(p)}{dp}\bigg|_{p^\star=1} = bp^{b-1}\Big|_{p^\star=1} = b,$$

implying,

$$\nu = \frac{\log b}{\log \frac{dR_b(p)}{dp}\Big|_{p^\star=1}} = 1,$$

so in 1d the renormalisation group transformation is exact.

Renormalisation Group Transformation on 2d Triangular Lattice

Divide the lattice into triangular cells containing three sites each. Let the probability for having a spanning cluster define the renormalisation group transformation. Thus,

$$R_b(p) = p^3 + 3p^2(1-p) = 3p^2 - 2p^3.$$

The fixed point equation,

$$R_b(p^\star) = 3p^{\star 2} - 2p^{\star 3} = p^\star \Leftrightarrow p^\star = \begin{cases} 0 \\ 1 \\ 1/2. \end{cases}$$

The two trivial fixed point is associated with the self-similar states of an empty lattice and fully occupied lattice. The unstable fixed point $p^\star = 1/2$ will be associated with the nontrivial self-similar states at p_c. The critical exponent ν is,

$$\nu = \frac{\log b}{\log \frac{dR_b(p)}{dp}\Big|_{p^\star=1/2}} = \frac{\log\sqrt{3}}{\log\left(6p-6p^2\right)\Big|_{p^\star=1/2}} = \frac{\log\sqrt{3}}{\log 3/2} = 1.355.$$

The exact values are $p_c = 1/2$ and $\nu = 4/3$ so the renormalisation group transformation does a good job.

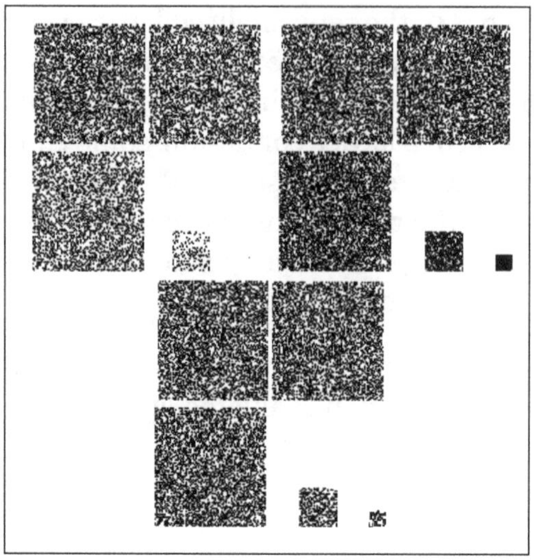

Figure above shows Renormalisation group method on a 2d square lattice according to the majority rule for 3 × 3 blocks. Shown are part of the original lattice L = 729 and the renormalised lattices L = 243 (part of), L = 81, 27, and 3. (a) Initial occupation probability $p < p^\star$ in the original lattice and the flow is towards the empty lattice. (b) Initial occupation probability $p > p^\star$ in the original lattice and the flow is towards the fully occupied lattice. (c) Initial occupation probability $p = p^\star$ in the original lattice. This is a fixed point for the renormalisation group transformation and consequently there is no flow. It looks like itself on all length scales.

Renormalisation Group Transformation on 2d Square Lattice of Bond Percolation

Consider the block defined in figure below which, after the coarse graining procedure, is replaced by two superbonds. The renormalisation group transformation is,

$$\begin{aligned} p' &= R_b(p) \\ &= \text{probability to have a spanning cluster in horizontal direction} \\ &= p^5 + p^4(1-p) + 4p^4(1-p) + 2p^3(1-p)^2 + 2p^3(1-p)^2 + 4p^3(1-p)^2 + 2p^2(1-p)^3 \\ &= 2p^5 - 5p^4 + 2p^3 + 2p^2 \end{aligned}$$

with the fixed point equation,

$$R_b(p^\star) = 2p^{\star 5} - 5p^{\star 4} + 2p^{\star 3} + 2p^{\star 2} = p^\star \Leftrightarrow p^\star = \begin{cases} 0 \\ 1 \\ 1/2. \end{cases}$$

Figure shows Renormalisation group method of 2d square lattice with bond percolation according to the spanning cluster rule. (a) and (b) After the coarse graining procedure, the block is replaced with two super bonds AC and AG. (c) The various configurations with a horizontal spanning cluster. The two dangling bonds DG and EH need not be considered as they do not affect the probability of having a spanning cluster from left to right.

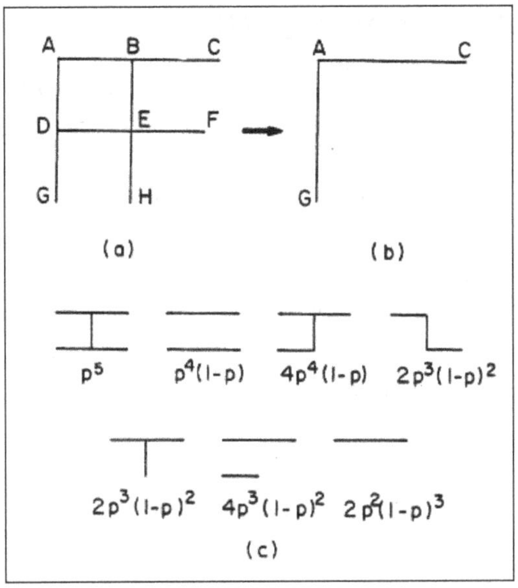

Again, the two trivial fixed points are associated with the self-similar states of an empty lattice and fully occupied lattice. The unstable fixed point $p^\star = 1/2$ will be associated with the nontrivial self-similar states at p_c. For bond percolation in 2d, $p_c = 1/2$. The critical exponent ν is,

$$\nu = \frac{\log b}{\log \left.\frac{dR_b(p)}{dp}\right|_{p^\star=1/2}} = \frac{\log 2}{\log 13/8} = 1.428,$$

which should be compared to the analytical result of $\nu = 4/3$ in 2d. Thus the renormalisation group transformation defined above for bond percoaltion in 2d results in an exact prediction of p_c and a good estimate of ν.

Gibbs Paradox and Indistinguishability of Particles

The expression obtained for entropy of a monoatomic ideal gas derived as given below:

$$S = Nk_B \ \text{In}\left[\left(\frac{V}{h^3}\right)\left(\frac{4\pi mE}{3N}\right)^{\frac{3}{2}}\right] + \frac{3}{2}Nk_B$$

Where, $E = \frac{3}{2}Nk_B T$. One of the interesting observations is that entropy as given in equation above, led to correct equation of state for an ideal gas and other thermodynamic properties. Despite these successes this expression lacks one of the most important properties of entropy called extensiveness which implies that entropy scales if we scale the extensive parameters involved in the expression. This implies $S(\alpha N,\alpha V,\alpha E) = \alpha S(N,E,V)$. In this module we seek to look for the implications of the lack of extensiveness of entropy S leading to Gibbs Paradox, and how this paradox can be resolved, bringing into focus the issue indistinguishability of particles involved.

Violation of Extensiveness of Entropy (S)

To check the extensiveness of entropy $S(N,E,V)$ given in equation,

$$S = Nk_B \text{ In}\left[\left(\frac{V}{h^3}\right)\left(\frac{4\pi mE}{3N}\right)^{\frac{3}{2}}\right] + \frac{3}{2}Nk_B,$$

Let us change $N \to \alpha N, V \to \alpha V$ and $E \to \alpha E$ keeping the intensive variables unchanged then equation $S = Nk_B \text{ In}\left[\left(\frac{V}{h^3}\right)\left(\frac{4\pi mE}{3N}\right)^{\frac{3}{2}}\right] + \frac{3}{2}Nk_B$ becomes,

$$S(\alpha N, \alpha V, \alpha E) = \alpha N k_B \text{ In}\left[\left(\frac{\alpha V}{h^3}\right)\left(\frac{4\pi m \cancel{\alpha} E}{3 \cancel{\alpha} N}\right)^{\frac{3}{2}}\right] + \frac{3}{2}N\alpha k_B \neq \alpha S(N,E,V)$$

Obviously, α sitting with V inside the argument of ln fails to make entropy extensive. This means that if system is made up of α parts each having entropy S then sum of the entropies of each part is ≠ α times S. This is physically unacceptable situation.

Entropy of Mixing of Two Ideal Gases

To fully see the implication of the violation of extensiveness of entropy, Gibbs went for a thought experiment of mixing of two samples of different ideal gases at the same temperature T, followed by mixing of two samples of same ideal gas at the same temperature. It is worth noting that first experiment involving mixing of different ideal gases is an irreversible process for which $\Delta S > o$, whereas second experiment mixing of two samples of same gas is a reversible process i.e. $\Delta S > o$.

Let us see these two case separately, and calculate explicitly using equation,

$$S = Nk_B \text{ In}\left[\left(\frac{V}{h^3}\right)\left(\frac{4\pi mE}{3N}\right)^{\frac{3}{2}}\right] + \frac{3}{2}Nk_B$$

Mixing of two samples of two different ideal gases at same temperature having same number density as shown in figure below shows the two different ideal gases 1 and 2 at the same temperature and having same number density i.e. $\left(\frac{N_1}{V_1} = \frac{N_2}{V_2}\right)$.

Entropy of N Three Dimensional Classical Harmonic Oscillators

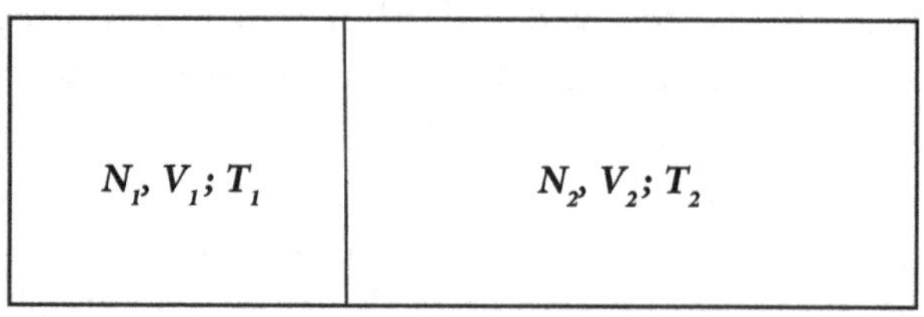

Mixing of two ideal gases.

Before mixing the entropy of two gases are respectively.

$$S_i = N_i k_B \text{In}\left[\left(\frac{V_i}{h^3}\right)(2\pi m_i k_B T)^{\frac{3}{2}}\right] + \frac{3}{2} N k_B \text{ where i = 1, 2}$$

Or

$$S_i = N_i k_B \text{In}\, v_1 + \frac{3}{2} N_i k_B \left[\text{In}\left(\frac{2\pi m_i k_B T}{h^2}\right) + 1\right] \text{where i = 1, 2}$$

After the mixing has taken place, the total entropy can be written easily keeping in mind that now the partition of figure has been removed and each particle of the gas can roam in total volume $V = V_1 + V_2$, so that total entropy can be written as,

$$S_{Total} = \sum_{1=1}^{z}\left[N_i k_B \text{In}\, V + \frac{3}{2} N_i k_B \left[\text{In}\left(\frac{2\pi m_i k_B T}{h^2}\right) + 1\right]\right]$$

Now we can define entropy of mixing as,

$$\Delta S = S_{total} - \sum_{i=1}^{z} S_i$$

Using $\left(S_i = N_i k_B \text{In}\, v_1 + \frac{3}{2} N_i k_B \left[\text{In}\left(\frac{2\pi m_i k_B T}{h^2}\right) + 1\right]\right)$ and $\left(S_{Total} = \sum_{1=1}^{z}\left[N_i k_B \text{In}\, V + \frac{3}{2} N_i k_B \left[\text{In}\left(\frac{2\pi m_i k_B T}{h^2}\right) + 1\right]\right]\right)$, this leads to,

$$\Delta S = k_B \left[N_1 \text{In}\frac{V}{V_1} + N_2 \text{In}\frac{V}{V_2}\right], \text{where } V = V_1 + V_2$$

Since $V > V_1$ and $V > V_2, \Delta S > 0$. Hence process is irreversible, expected for mixing. Furthermore, since we assumed that number density is same, equation above can be written as,

$$\Delta S = k_B \left[N_1 \text{In}\frac{N_1 + N_2}{N_1} + N_2 \text{In}\frac{N_1 + N_2}{V_2}\right]$$

Once again $\Delta S > 0$.

Mixing of two samples of same ideal gases at same temperature having same number density.

Now let us look at the same mixing process by taking in the two compartments same gas and remove the partition allowing for mixing the gases of the two compartments. Only difference now is that $m_1 = m_2 = m$. So that S_{total} becomes,

$$S_{total} = N k_B \text{In}\, V + \frac{3}{2} N k_B \left[\text{In}\left(\frac{2\pi m k_B T}{h^2}\right) + 1\right] \text{where } N = N_1 + N_2$$

Using this expression, and the expression the entropy of mixing again remains the same as given in equation $\Delta S = k_B\left[N_1 \text{In}\frac{N_1+N_2}{N_1}+N_2 \text{In}\frac{N_1+N_2}{V_2}\right]$ with $\Delta S > 0$. This is where the paradox arises, mixing of two samples of the same gas again when temperature T and Number density are same can be undone merely by inserting the partition, making the whole process reversible and the entropy of mixing ΔS should be equal to zero, which is not the case. In literature this is known as Gibbs Paradox.

Adhoc Resolution of Gibbs Paradox

As discussed before, there is a need to give a careful look to equation

$S = Nk_B \text{ In}\left[\left(\frac{V}{h^3}\right)\left(\frac{4\pi mE}{3N}\right)^{\frac{3}{2}}\right]+\frac{3}{2}Nk_B$ so that entropy does not remain nonextensive and paradox of entropy of mixing ΔS of two samples of same gas does not become greater than zero for a reversible process. Gibbs proposed an adhoc way of overcoming this paradox by suggesting that from equation $S = Nk_B \text{ In}\left[\left(\frac{V}{h^3}\right)\left(\frac{4\pi mE}{3N}\right)^{\frac{3}{2}}\right]+\frac{3}{2}Nk_B$ if we subtract $\text{In}\, N!$. the entropy of mixing for becomes zero and restores extensiveness of the entropy of a nonoatomic ideal gas. This is equivalent to diminishing S_1 by $\text{In}\, N_1!$, S_2 by $\text{In}\, N_2!$ and S by $\text{In}\, N!$

If we do this equation $S = Nk_B \text{ In}\left[\left(\frac{V}{h^3}\right)\left(\frac{4\pi mE}{3N}\right)^{\frac{3}{2}}\right]+\frac{3}{2}Nk_B$ becomes,

$$S = Nk_B \text{ In}\left[\left(\frac{V}{h^3}\right)\left(\frac{4\pi mE}{3N}\right)^{\frac{3}{2}}\right]+\frac{3}{2}Nk_B - k_B \text{ In } N!$$

Or

$$S = Nk_B \text{ In}\left[\left(\frac{V}{h^3}\right)\left(\frac{4\pi mE}{3N}\right)^{\frac{3}{2}}\right]+\frac{3}{2}Nk_B - k_B(N \text{ In } N - N)$$

Or

$$S = Nk_B \text{ In}\left[\left(\frac{V}{N}\right)\left(\frac{4\pi mE}{3Nh^3}\right)^{\frac{3}{2}}\right]+\frac{5}{2}Nk_B$$

And hence energy E, after simplification, can be written as,

$$E = \frac{3h^2N^{\frac{5}{3}}}{4\pi mV^{\frac{2}{3}}}\exp\left(\frac{2S}{3Nk}-\frac{5}{3}\right)$$

A mere glance tells us that entropy has become extensive with a factor of N dividing V inside the logarithmic term.

Let us see introducing this adhoc term physically amounts to doing what. This amounts to diminishing the number of microstates of the monoatomic ideal gas by a factor of N! involving new counting procedure called correct Boltzmann Counting Process. If we recall problems which we have done in permutations and combinations which allow us to count without counting, the number of permutations of N different things taken N at a time is given by N!, but if all the N things are identical it should lead to 1 possibility only this requires us to divide N! by N!.

So while checking the extensiveness of the equation $S = Nk_B \ \text{In}\left[\left(\frac{V}{h^3}\right)\left(\frac{4\pi mE}{3N}\right)^{\frac{3}{2}}\right] + \frac{3}{2}Nk_B$, therefore, need is to revisit the process of counting of microstates of a monoatomic ideal gas and division by N! an adhoc solution to Gibbs paradox amounts to saying that though we are dealing with the classical particles constituting the ideal gas still these should be treated as identical to get physically correct entropy. This expression for entropy of an ideal gas was first calculated by O. Sackur and H. Tetrode in a series of papers separately during 1911-12 using for the first time quantization of phase space in classical statistical mechanics.

Chemical Potential

Chemical potential is defined as $\mu = \left(\frac{\partial E}{\partial N}\right)_{V,S,}$ therefore,

$$\mu = \frac{3h^2N^{\frac{5}{3}}}{4\pi mV^{2/3}}\left[\frac{5}{3N} - \frac{2S}{3N^2k_B}\right]\exp\left(\frac{2S}{3Nk_B} - \frac{5}{3}\right)$$

Or

$$\mu = \left[\frac{5}{3N} - \frac{2S}{3N^2k_B}\right]E$$

In terms of N, V, T, equation above can be written as,

$$\mu = k_B T \ \text{In}\left[\frac{N}{V}\left(\frac{h^2}{2\pi mk_BT}\right)^{\frac{3}{2}}\right]$$

It is interesting to note that in classical limit, when particle density is low and temperature very high i.e. $\frac{N}{V} \ll \left(\frac{h^2}{2\pi mk_BT}\right)^{\frac{3}{2}}$. This implies that the argument of logarithmic term is always less than 1 and hence chemical potential is always negative. The quantity $\frac{(2\pi mk_BT)^{\frac{1}{2}}}{h}$ is called thermal de Broglie wavelength.

Gibbs Free Energy

Gibbs free energy is defined as $G=\mu N$,

$$G=Nk_BT\ln\left[\frac{N}{V}\left(\frac{h^2}{2\pi mk_BT}\right)^{\frac{3}{2}}\right]$$

Helmholtz Free Energy

Helmholtz free energy has been defined as $F=E-TS=G-PV=\mu N-Nk_BT$,

$$F=Nk_BT\left[\ln\left[\frac{N}{V}\left(\frac{h^2}{2\pi mk_BT}\right)^{\frac{3}{2}}\right]-1\right]$$

Indistinguishability and Enumeration of Microstates

Indistinguishability of particles in a classical ideal gas means that by no means you can tag these particles as particle A, particle B, particle C, and so on. At the most we can talk about is the distribution of these indistinguishable particles over different states, say n_i particles over the state ε_i, giving a set of number of particles $\{n_i\}$ distributed over a set of states $\{\varepsilon_i\}$ such that $\sum_i n_i = N$ and $\sum_i n_i\varepsilon_i = E$.

We shall carefully look at the emerging situation, the total number of permutations of N particles distributed according to the set $\{n_i\}$ is like distributing N particles such that n_1 particles are of the kind sitting in state ε_1 ,n_2 particles are of the kind sitting in state ε_2 and so on. Thus the total number of permutations turn out to be,

$$\frac{N!}{n_1!n_2!\ldots\ldots\ldots\ldots}$$

Where, the set $\{n_i\}$ satisfies the two constraints corresponding to total number of particles and total energy stated above. If all the particles had been distinguishable this would have given us the correct number of microstates of a given macrostate (N,V,E). But because particles are indistinguishable, each of these microstates gives one and only one microstate. So Gibbs remedy dividing by N! still leaves us with a spurious weight factor,

$$\frac{1}{n_1!n_2!\ldots\ldots\ldots\ldots}$$

The only way to make it 1 is probability of each of n_i greater than 1 becomes smaller and smaller and most of the time it takes values either 0 or 1and rarely greater than 1. This will bring the expectation value of $n_i \ll 1$. which defines the classical limit. This leads to correct enumeration of microstates of a given macrostate, which, however, can be understood only by quantum statistics and not otherwise.

Friction and Fluctuations

Friction and Dissipation

Friction is a familiar fact of life. Rub two objects together and you feel a force resisting the motion. Place an object on the floor and give it a push; it will slide a short distance and then stop. Again, there is a force resisting its motion: friction. And something else happens too: the objects moving against each other become warmer, as when you rub your hands to warm them up or when you strike a match. The same thing happens in other contexts too. For example, electrical resistance is really just another type of friction. If you run a current through a conductor, a force resists the motion of the electrons, and the conductor becomes warmer.

Friction is such a universal part of our experience, it seems like it must reflect something fundamental in the laws of nature. Yet if you look at the basic equations of classical or quantum mechanics, there is no sign of friction. A single isolated particle will continue forever moving in a straight line without slowing down. If two particles collide, they bounce away from each other with just as much energy as they started with. Friction, it appears, only happens to macroscopic objects. And there are even situations where macroscopic systems are free from it, such as superconductivity and superfluidity. What exactly is friction, and where does it come from?

The answer, of course, is statistical mechanics. When a system is in equilibrium, its kinetic energy is uniformly distributed among all its degrees of freedom. A macroscopic object in motion clearly is *not* in equilibrium. It has one degree of freedom (the center of mass motion) with far more energy than any other. If that degree of freedom interacts with others, the system will tend to move toward equilibrium. Energy will be transfered out of that one degree of freedom (so the object will slow down) and into all the others (so the object will become warmer).

That is all friction really is: the tendency of systems to move toward equilibrium when different degrees of freedom are allowed to interact with each other. In the process, its energy becomes less useful. The kinetic energy of a moving macroscopic object is useful energy. It can easily be used to do mechanical work. Thermal energy is less useful. You need a heat engine with a separate cold bath to do work with it, and Carnot's theorem sets a strict upper limit on how much work it can do. This conversion of useful mechanical energy to less useful thermal energy is called dissipation. We say that friction "dissipates energy". That is, it converts mechanical energy to thermal energy.

Linear Response Theory

Even when we studied about thermodynamic processes, we only considered transitions between two equilibrium states. We assumed that if we waited long enough, the system would eventually return to equilibrium, but we said nothing about how that process took place. Friction *is* that process. A system experiencing friction is, by definition, a system out of equilibrium.

When a system is in equilibrium, all mechanical and thermodynamic forces exactly cancel out. Its macroscopic variables tend to remain fixed. The converse is also true: if a system is *not* in equilibrium, the mechanical and thermodynamic forces will in general *not* cancel out, and the system will experience a net force pushing it back toward equilibrium. Intuitively, we expect that the further the system is from equilibrium, the stronger the force will be. But how exactly does it vary? We are simply going to make an assumption:

Assumption (Linear Response)

If a macroscopic variable is displaced away from equilibrium by a distance Δx, the system will experience a net restoring force that is linearly proportional to the displacement: $F = -C\Delta x$, where C is a constant.

This assumption is the basis of *Linear Response Theory.* The macroscopic variable could be the velocity of a pollen grain immersed in water, the volume of an inflated balloon, or the electric current traveling through a circuit. All that matters is that it interacts with many microscopic degrees of freedom that can act as a heat bath.

How do we justify this assumption? If the system is not in equlibrium, the net force will almost certainly tend to push it back toward equilibrium. For example, the pollen grain will collide with water molecules that slow it down. Or the balloon will be bombarded by air molecules that apply pressure to it. But there is no fundamental reason the force must be linear in the displacement.

We will justify it in two ways. First, suppose the force is not linear in the displacement. In that case, we can expand it in a power series about the equilibrium value. If Δx is sufficiently small, all terms after the linear one will be negligible. So even though the assumption is not strictly correct, it is still a good approximation as long as the displacements are small enough. The second justification is simply an experimental observation: linear response turns out to be a very good description of many important systems.

Fluctuations

Friction is caused by the interaction with a heat bath. Let us be very clear about that. A moving object slows down because it is colliding with molecules in its environment. In the process, energy is dissipated: the object's kinetic energy is redistributed among a huge number of microscopic degrees of freedom. These two phenomena, friction and dissipation, are really exactly the same thing.

But that is not the only effect of being in contact with a heat bath. Suppose a pollen grain is initially not moving at all. In that case, collisions with water molecules will cause it to start moving. In equilibrium, after all, every degree of freedom should have $kT/2$ kinetic energy. If it starts with less than that, the effect of the heat bath will be to increase its kinetic energy.

So in equilibrium, the pollen grain will be in constant motion. Its speed and direction of motion will be constantly changing, as it is jostled by the surrounding water molecules. But on average, its kinetic energy will equal the value given by the equipartition theorem.

We refer to this effect as equilibrium fluctuations. Any variable in equilibrium with a heat bath will be constantly changing. The changes will appear random, but they will obey statistical rules determined by the nature of the heat bath.

These two effects, fluctuations and dissipation, are inseparable from each other. They are both caused by exactly the same mechanism: the interaction of the variable with a heat bath. And because they have the same cause, they are guaranteed to obey certain relationships. All that remains is to derive what those relationships are.

The Langevin Equation

Let's write Newton's Second Law for a particle undergoing Brownian motion in water: $F = m\ddot{x}$. (For simplicity we will work in one dimension, but the generalization to more dimensions is trivial. Just turn x and F into vectors.) In this equation, F is the force exerted on the particle by the surrounding water molecules. The details of those interactions are enormously complicated and constantly changing, so we need to look for an approximate description of it. What would be a reasonable approximation?

The first thing we might try is using the assumption of linear response, so the force at any moment is proportional to the velocity: $m\ddot{x} = -\gamma\dot{x}$. The parameter γ is called the friction coefficient. The solution to this differential equation is given by,

$$\dot{x}(t) = \dot{x}(0) \cdot e^{-\frac{\gamma}{m}t}$$

For a truly macroscopic system, this description might be sufficient. Whatever velocity the object starts with, it exponentially decays toward zero. But Brownian motion is not precisely a macroscopic phenomenon. It applies to particles that are much larger than a water molecule, but still small enough that you need a microscope to see them. And the whole point is that their velocity does *not* decay to zero. They remain in motion due to the constant collisions with water molecules. By ignoring the forces applied by those random collisions, we have simplified things too far.

We can still assume linear response, but now we will only take it as a statement about the *average* force on the particle. At any given instant, the force could be different. This leads to the following equation, known as the *Langevin equation*:

$$m\ddot{x} = -\gamma\dot{x} + R$$

R is a "random" force describing the rapidly fluctuating interactions between the particle and water molecules. We cannot hope to write an exact function for it, but we can still describe it statistically. We will make the following assumptions about it:

1. $\langle R \rangle = 0$. We already separated out the average force into its own term, so R must have a mean of 0.

2. It is independent of x. The interaction between the particle and the water is the same no matter where in the water bath the particle is located.

3. Because it varies so rapidly and chaotically, we assume it is uncorrelated with itself except over very short time spans. More precisely, we assume there is some maximum time τ over which it has any correlations, so that,

   $$\langle R(t)R(t-\delta t)\rangle = 0 \text{ if } \delta t > \tau$$

4. We also assume the rate at which correlations decay is independent of time, so that $\langle R(t)$

R(t + δt)⟩ is independent of t. It depends only on δt. The overall statistical properties of the random force are the same at all times.

These assumptions have an important consequence. In most situations, we do not care about the instantaneous value of R, only its integral over some time period that is long compared to τ. We can break up that integral into many pieces, each covering a span of length τ:

$$\int_0^t R(t')dt' = \int_0^\tau R(t')dt' + \int_\tau^{2\tau} R(t')dt' + \int_{3\tau}^{3\tau} R(t')dt' + \ldots$$

So the integral is a sum of independent terms, each drawn from the same distribution. So we can immediately apply the central limit theorem and conclude that the integral obeys a normal distribution with mean 0 (because ⟨R⟩=0), and whose standard deviation scales with $\sqrt{t}$. All that, without evaluating a single integral or knowing anything about the details of R(t).

Before we dive into the math, let's take a moment to look at the Langevin equation and try to understand it intuitively. There are two terms on the right side. The first one always points opposite to the velocity, so it tends to slow the particle down. If it were the only term, it would lead to exponentially decaying velocity as in equation $\dot{x}(t) = \dot{x}(0) \cdot e^{-\frac{\gamma}{m}t}$. The second term prevents that from happening by constantly applying random kicks to the particle. Essentially, there is one term that removes energy and one term that adds energy. When the system is in equilibrium, the two terms will exactly balance out so the average energy remains constant. The magnitude of the random force will of course depend on temperature: the hotter the system, the faster the water molecules will be moving, and the harder they will hit the particle. We therefore expect there should be some relationship between the temperature, the friction coefficient, and the magnitude of the random force.

Diffusion of a Brownian Particle

Now that we have the pieces in place, it is time to see what we can calculate. To start with, multiply both sides of equation $m\ddot{x} = -\gamma\dot{x} + R$ by x, then take the ensemble average:

$$m\langle x\ddot{x}\rangle = -\gamma\langle x\dot{x}\rangle + \langle xR\rangle$$

Because R is independent of x, the last term vanishes: ⟨xR⟩=⟨x⟩⟨R⟩ = 0. To unify the other two terms, we can use the identity,

$$\frac{d}{dt}\langle x\dot{x}\rangle = \langle x\ddot{x}\rangle + \langle \dot{x}^2\rangle$$

(This identity makes use of the fact that the operations of taking a derivative and taking an ensemble average commute with each other. After all, an ensemble average is just a weighted sum of terms that can each be differentiated independently.) Substituting this into equation $m\langle x\ddot{x}\rangle = -\gamma\langle x\dot{x}\rangle + \langle xR\rangle$ gives,

$$m\left(\frac{d}{dt}\langle x\dot{x}\rangle - \langle \dot{x}^2\rangle\right) = -\gamma\langle x\dot{x}\rangle$$

Next we can use the equipartition theorem. A single particle in one dimension has one degree of freedom so, in equilibrium, its average kinetic energy is $m\langle \dot{x}^2 \rangle / 2 = kT/2$.

Therefore,

$$m\frac{d}{dt}\langle x\dot{x}\rangle = kT - \gamma\langle x\dot{x}\rangle$$

Now make a change of variables. Define,

$$z = \frac{d}{dt}\langle x^2 \rangle = 2\langle x\dot{x}\rangle$$

so the equation becomes,

$$\frac{dz}{dt} = \frac{2kT}{m} - \frac{\gamma}{m}z.$$

This is a standard differential equation whose solution is given by,

$$z = Ce^{-\frac{\gamma}{m}t} + \frac{2kT}{\gamma}$$

You can easily verify that by substituting it back into equation $\frac{dz}{dt} = \frac{2kT}{m} - \frac{\gamma}{m}z$. C is an arbitrary constant.

We are almost done! The first term in equation $z = Ce^{-\frac{\gamma}{m}t} + \frac{2kT}{\gamma}$ decays rapidly with time. Taking the limit of large t, we get,

$$z = \frac{d}{dt}\langle x^2 \rangle = \frac{2kT}{\gamma}$$

or integrating with respect to time and defining x(0)=0,

$$\langle x^2 \rangle = \frac{2kT}{\gamma}t$$

So as time passes, the mean squared distance traveled by the particle increases at a constant rate. This result is not really very surprising. In fact, it is exactly what we probably would have predicted if we had thought about it in advance. The particle moves irregularly as its velocity is continuously damped away by friction and replaced by the random force. The motion can be thought of as a series of independent displacements taken at random, one after another. By the central limit theorem, they should therefore add up to a normal distribution whose standard deviation grows as $\sqrt{t}$. This kind of motion is called diffusion. It also is sometimes referred to as a random walk, because of the way it is built up from many independent random displacements.

The way it scales with temperature is similarly unsurprising. If we had been forced to predict in advance, we probably would have guessed the average displacement would be proportional to the average velocity, and by the equipartition theorem that scales as $\sqrt{T}$. We also would probably have guessed that a larger friction coefficient would lead to a smaller average displacement.

On the other hand, there is one strange thing about equation $\langle x \rangle \ \frac{2kT}{}t$. It appears to be completely independent of the random force. In fact, R dropped out of the derivation after the very first step and never appeared again. But the random force is what causes the particle to diffuse! Surely the average displacement ought to depend on the magnitude of R?

In fact, this result does depend on R, but it does it indirectly. By using the equipartition theorem, we assumed the system was in equilibrium at temperature T. That equilibrium is brought about by the interaction between the friction force and the random force, and R depends on T. To derive equation $\langle x^2 \rangle = \frac{2kT}{\gamma} t$, it was sufficient to assume equilibrium had been achieved without knowing precisely how it came about. So now we will turn to that question: how do the forces interact to produce equilibrium?

The Fluctuation-Dissipation Theorem

Equation $m\ddot{x} = -\gamma\dot{x} + R$ can be viewed as a first order differential equation for the velocity $\dot{x}$. Its solution is given by,

$$\dot{x}(t) = \dot{x}(0) \cdot e^{-\frac{\gamma}{m}t} + \frac{1}{m}\int_0^t e^{-\frac{\gamma}{m}(t-t')} R(t')dt'$$

The first term is identical to equation $\dot{x}(t) = \dot{x}(0) \cdot e^{-\frac{\gamma}{m}t}$. It represents the initial velocity decaying exponentially to zero. But this time, we also have a second term representing the effect of the random force. It can be viewed as a series of infinitesimal kicks, each of which then decays with time. The integral sums over all those kicks, weighting each one by an exponential factor based on the time $t - t'$ that has passed since it occurred.

We are interested in the behaviour at large values of t, once the system has had plenty of time to come to equilibrium. In this limit equation above becomes simpler:

$$\lim_{t \to \dot{x}(t)} = \frac{1}{m}\int_0^\infty e^{-\frac{\gamma}{m}(t-t')} R(t')dt'$$

Notice that t still appears in the exponential. You might be tempted to replace it with ∞, but that would be incorrect. As t increases, the upper limit of the domain of integration increases as well. The exponent only depends on the difference $t - t'$, and no matter how large t gets, there is always a portion of the domain of integration for which the exponent is small.

We want to compute the average kinetic energy in equilibrium, so square each side and take an ensemble average.

$$\lim_{t \to \infty} \langle \dot{x}^2(t) \rangle = \frac{1}{m^2}\int_0^\infty \int_0^\infty e^{-\frac{\gamma}{m}(2t-t'-t'')} \langle R(t')R(t'') \rangle dt'dt''$$

We assumed that $\langle R(t')R(t'')\rangle$ was independent of t' and depended only on the difference $t''-t'$. We can therefore replace it by $\langle R(0)R(t''-t')\rangle$. Also make a change of variables by defining $r=t-t''$ and $s=t''-t'$. With these substitutions, the equation becomes,

$$\lim_{t\to\infty}\langle \dot{x}^2(t)\rangle = \frac{1}{m^2}\int_{r=0}^{\infty}\int_{s=-\infty}^{\infty} e^{-\frac{\gamma}{m}(2r+s)}\langle R(0)R(s)\rangle drds$$

$$=\frac{1}{m^2}\left(\int_0^{\infty} e^{-\frac{\gamma}{m}2r}dr\right)\left(\int_{-\infty}^{\infty} e^{-\frac{\gamma}{m}s}\langle R(0)R(s)\rangle ds\right)$$

$$=\frac{1}{2\gamma m}\int_{-\infty}^{\infty} e^{-\frac{\gamma}{m}s}\langle R(0)R(s)\rangle ds$$

Now we can invoke the equipartition theorem once again to set $\langle \dot{x}^2(t)\rangle = kT/m$. Therefore,

$$2\gamma kT = \int_{-\infty}^{\infty} e^{-\frac{\gamma}{m}s}\langle R(0)R(s)\rangle ds$$

We assumed the random force was uncorrelated with itself except over some "very short" time τ. Let's make that assumption more precise. From equation $\dot{x}(t)=\dot{x}(0)\cdot e^{-\frac{\gamma}{m}t}+\frac{1}{m}\int_0^t e^{-\frac{\gamma}{m}(t-t')}R(t')dt'$, we see that m/γ sets the timescale on which friction operates. After a time m/γ has passed, the initial velocity has decayed by a factor of e. We will assume that $\tau \ll m/\gamma$. The timescale over which the random force is correlated with itself is very small compared to the timescale over which friction operates.

This assumption means that whenever the integrand is non-zero, the exponent is negligible in magnitude. This leads to a further simplification:

$$2\gamma kT = \int_{-\infty}^{\infty}\langle R(0)R(s)\rangle ds$$

This very important result is called the fluctuation-dissipation theorem. It gives the relationship between T, γ, and R. Any system in contact with a heat bath will experience both friction and fluctuations, and the magnitudes of the two effects are directly linked to each other. You cannot get one without the other.

The right side of equation above is the integral of the autocorrelation function of the random force. It depends both on the magnitude of R and on the time τ over which it remains correlated. The stronger the force is, the larger its effect on the particle. And the longer the time interval over which the particle accelerates in a single direction before the force changes, the faster it will get moving.

References

- Statistical-mechanics, science: britannica.com, Retrieved 29, June 2020
- IsingModel: uci.edu, Retrieved 10, March 2020
- Heisenberg-model-(quantum): wikiwand.com, Retrieved 27, August 2020
- Percolation: mit.edu, Retrieved 03, February 2020
- Friction: stanford.edu, Retrieved 18, April 2020

2

Statistical Thermodynamics

Statistical thermodynamics refers to a theory that predicts the behavior of macroscopic quantities of compounds by analyzing their molecular properties. Some of its important aspects are thermodynamic equilibrium, thermodynamic potentials and statistical foundation of thermodynamics. The topics elaborated in this chapter will help in gaining a better perspective about these facets of statistical thermodynamics.

Statistical mechanics grew out of an earlier field called thermodynamics, which was concerned with the thermal properties of liquids and gasses. It grew up around it, and then subsumed it. What we now call "classical thermodynamics" was developed over a period of several hundred years, but much of the most important work was done in just a few decades from the 1820s through the 1850s. It is not at all a coincidence, of course, that this burst of activity coincided with the industrial revolution and the development of the locomotive. Classical thermodynamics was largely developed by people who wanted to learn how to make better steam engines.

Statistical mechanics has come a long way from these humble beginnings, but thermodynamics is still an important field in its own right. Here, we will discuss some of the most important results of classical thermodynamics as seen from a modern statistical viewpoint.

Thermodynamic Processes

In everything that has come so far, we have considered only systems that were in equilibrium. We will now take a first step toward removing this restriction. Consider some physical system, and assume it is initially in equilibrium. We now change the system in some way. There are many types of changes we might make. Here are some examples:

- We could directly add energy to the system, for example by shaking it or by placing it over a fire.
- We could bring it into contact with a new heat bath. This might be done by moving it, or by opening a door that previously insulated it and prevented heat from flowing between it and the heat bath.
- We could change the volume of the system. For example, it might be contained in a chamber, one wall of which is attached to a piston allowing it to be moved in or out.
- We could change the potential energy function for the system. For example, we could apply an electric or magnetic field. This will change the energy of each microstate.

Once we make any of these changes, the system will no longer be in equilibrium. If we let it sit long enough, however, it will eventually come back into equilibrium. How long that takes will depend entirely on the system. It could easily be anything from microseconds to millions of years. Assume we do that, so we once again are dealing with a system in equilibrium. This whole process of changing the system and allowing it to equilibrate is known as a thermodynamic process. We can now compare the two different states of the system, one before the process was performed and one after.

We have fundamentally changed our definition of what can constitute a physical system. Up to now we have dealt with two types of systems:

- An isolated system whose macroscopic properties, like energy and volume, are fixed and unchanging.
- A system in contact with a heat bath. The properties of the system are determined by the properties of the heat bath. They are free to change, but they do so at random and mostly only within a narrow range. Also, if we view the system of interest plus the heat bath as a single composite system, that composite system is still isolated.

We now have a third option: a system whose properties we can directly control. They change when we choose to change them in exactly the way we choose. We are no longer in any way dealing with an isolated system. We are explicitly making an external change, reaching in from outside to alter the system. On the other hand, we only do that during a restricted time period. Before we make the change, the system (possibly including a heat bath) is isolated. After we are done making the change it is again isolated. But while we make the change, it is not isolated.

The Laws of Thermodynamics

Classical thermodynamics is based on four laws that are very sensibly numbered starting at zero. When viewed from a modern perspective, three of them are trivial. The remaining one is extremely profound, and even today is often misunderstood.

The Zeroth Law of Thermodynamics: If two bodies are each in thermal equilibrium with a third body, then they are also in thermal equilibrium with each other. This law was needed to justify that one can define temperature in terms of values measured on a thermometer. Remember that when thermodynamics was first being developed, no one was certain exactly what "temperature" really was. It was strictly an empirical measurement. This law asserts that if you let a thermometer come to equilibrium with two different bodies, and they each produce the same measurement, then bringing those bodies together will produce no net flow of heat.

Once we associate temperature with the average kinetic energy per degree of freedom, this law becomes trivial. It is merely the statement that if two numbers each equal a third number, then they also equal each other.

The First Law of Thermodynamics: The change in a system's energy during a thermodynamic process equals the heat added to the system, minus the work it performs on its environment. In differential form, this is written,

$$dE = dQ - dW$$

where Q represents heat and W represents work.

This is just a statement of conservation of energy. But once again, remember that the founders of classical thermodynamics did not know exactly what "heat" was. The fact that it was a form of energy, and that it participated in the conservation of energy, had to be discovered through experiment.

Also notice how the form in which it is stated reflects the sorts of problems they were trying to study: mechanical engines that one adds heat to and that do work. In many cases, it is hard to cleanly divide all energy conversions into "heat" and "work", for example in processes that involve chemical reactions or in which matter is transferred between the system and a heat bath. But of course, energy conservation is a completely general law and applies to all situations equally well. The next law can be stated in several forms which appear different, but are actually equivalent. Here are two of the most common versions.

The Second Law of Thermodynamics: In any thermodynamic process, the total entropy either increases or remains constant, but never decreases or heat will never spontaneously flow from a colder body to a warmer one. This law reflects a deep truth about the nature of the universe we live in. It deserves a whole section of its own, so before I get to that let me first finish off the list of laws.

The Third Law of Thermodynamics: As the temperature of any system approaches zero, its entropy approaches a minimum value. This is another law that was needed to allow empirical definitions of temperature and entropy. From a statistical viewpoint, it is a statement of the fact that for nearly all physical systems, the density of states increases with energy.

The Second Law of Thermodynamics

The second law is not a trivial consequence of statistics or mechanics, but at first glance it is easy to *think* it is. The entropy of a system is simply the logarithm of its density of states. For an isolated system, that is a measure of how probable each macrostate is. The second law could therefore be summarized as, "Systems tend to move from less probable states to more probable ones." That sounds obvious. If a system is initially in a less probable state, we should hardly be surprised to later find it has moved to a more probable one. If it went the other way and moved to a less probable state, *that* would be surprising.

Suppose you have a box filled with gas molecules. Now you remove the cover, so the molecules are free to leave the box. Clearly it is far more likely for the molecules to spread out through the room than to all remain inside the box. There are far more possible arrangements of them when they are spread out through the room. That is a higher entropy state. Once they have diffused out into the room, it is incredibly unlikely you will ever find that, simply by chance, they are all inside the box at the same time again. That is what the second law tells us.

But there is a problem with this logic. So far as we know, the laws of physics are symmetric with respect to time. (Or more accurately they obey CPT invariance, but that distinction is irrelevant for this discussion.) If you reverse the velocity of every particle, the entire system will exactly retrace its trajectory in reverse. There is nothing in the laws of physics that distinguishes between "forward" and "backward" in time. Those are just arbitrary labels we assign to two opposite directions. If we reversed the labels, nothing about the laws of physics would change. (For CPT invariance, you also need to reverse the labels "left" and "right", as well as the labels "positive" and "negative" charge.)

So we can repeat the same argument in reverse. Consider a room containing gas molecules and an open box. Now close the box. Assume that after you close it, all the molecules are inside the box.

At any earlier point in time when the box was still open, it was incredibly unlikely that all the molecules just happened to be inside the box. So this "proves" that entropy tends to decrease.

Boltzmann himself spent years trying to prove the second law, and several times thought he had succeeded. The most famous of these attempts is known as the "H-theorem", published in 1872. It appeared to be a completely rigorous proof of the second law for a gas of classical particles. Of course it contained an assumption that broke time symmetry, but that assumption came into it in such a subtle way that for a few years neither Boltzmann nor any of his contemporaries noticed it. Then in 1876, Josef Loschmidt published his "reversibility objection" to the H-theorem. He reasoned as follows:

Start with a system in a low entropy (non-equilibrium) state. Given enough time, it will eventually move to a higher entropy, equilibrium state. Now reverse the velocity of every particle. Since the equations of motion are time symmetric, the system will retrace its trajectory and return to the original, low entropy state (except with velocities reversed). Clearly the second law cannot be a universal law that applies in all cases, because we have just presented a counter-example: a case where the entropy spontaneously decreases.

In response to this objection, Boltzmann slightly modified his position, arguing the second law was still true in a probabilistic sense. He acknowledged that one could carefully design states whose entropy would decrease. But any state that was not specifically designed in this way would almost certainly be one whose entropy increased rather than decreasing.

Unfortunately, this claim is simply wrong. The original low entropy state, the high entropy state it evolves into, and the time reversed version of that high entropy state (which will spontaneously return to a lower entropy one) are all equally probable. There is no basis for accepting some while ignoring others. For every trajectory whose entropy increases, there is a reversed trajectory whose entropy decreases. Both directions are equally probable.

Later in his career, Boltzmann finally came to a clearer understanding of the situation. He recognized that the second law *cannot* be proven. No amount of statistics or probability can turn a symmetric theory into a non-symmetric one. The second law of thermodynamics is not really a "law" at all. It is a *definition*. When we speak of "forward in time," we really mean "the direction of increasing entropy." All the phenomena that lead us to view time as asymmetric follow directly from the change in entropy: that it is easy to break a window but hard to fix it; that friction makes things slow down instead of speeding up; even that you can remember the past but not the future. If the arrow of time were reversed so that entropy increased in the opposite direction, you would not notice any difference at all. You would simply perceive the opposite direction as being "forward in time."

Boltzmann was unable to explain *why* entropy is changing in the first place. If the universe were in equilibrium, entropy should be constant and unchanging in all directions. The best answer he could suggest was an anthropic argument. He observed that even in equilibrium, highly improbable states can still happen. In fact, if you wait long enough, *every* state will eventually be visited. The entire Earth could be a random fluctuation that occurred just by chance. A fluctuation of that sort is incredibly improbable; most of the time, the atoms making up the Earth should be arranged in a generally uniform, featureless way. But in that situation, life could not exist. Given that we *do* exist, we should not be surprised to find the universe in one of the rare states that can support life.

This was a very clever idea, and in many ways a very modern one. It was, however, wrong. It leads to some very specific predictions, and those predictions do not match the universe we actually see around us.

Today we have a better explanation. The change in entropy is an effect of the Big Bang. We know that roughly 13.8 billion years ago, the universe was in a state of incredibly low entropy with all its matter and energy concentrated in a tiny region of space. Entropy has been steadily increasing ever since, and will continue increasing far into the future. That is why we find ourselves in a region of spacetime where there is an entropy gradient. For practical purposes, the phrase "forward in time" really means, "away from the Big Bang."

We still do not know *why* the Big Bang happened. There are various ideas about that, but we do not yet have enough evidence to say which one is correct. There may be other regions of spacetime where entropy is flat and unchanging. If so, those regions cannot support life, so it is not surprising we are not in one of them. We do not know whether the Big Bang marks the boundary of the universe, or whether there is more universe on the other side of it. If the latter, then entropy in that region presumably increases in the opposite direction. Anyone living there would perceive time as moving in the opposite direction to how we perceive it. Perhaps there are many different Big Bangs and many different universes surrounding them, each with its own arrow of time.

Three Iconic Equations in Thermodynamics

Consider three iconic equations encountered in the education of a physicist:

1. Thermodynamic entropy: In 1862 Clausius proposed that entropy is,

$$\Delta S = \int \frac{dQ_{rev}}{T}$$

where dQ is heat transferred during reversible heat flow (at constant volume). Heat is energy that spontaneously flows when objects of different temperatures are placed in thermal contact.

2. Maxwell-Boltzmann distribution for the ideal gas: In 1860 Maxwell produced a symmetry argument that eventually led to a form of the Maxwell-Boltzmann distribution equation for of the speed (not velocity) of atoms in an ideal gas:

$$f_{|\upsilon|}(\upsilon) = A\upsilon^2 \exp\left(-\frac{m\upsilon^2}{2k_B T}\right)$$

where, $A = 4\pi\sqrt{\left(\frac{m}{2\pi k_B}\right)^3}$ is a normalization constant, and k_B (or k) is Boltzmann›s constant, k_B, equals 1.38062 x 10−23 J/K.

Statistical Definition of Entropy

$$S = k_B \ln \Omega$$

S=k. log W

The meaning of *W* (here called Ω is still debated.

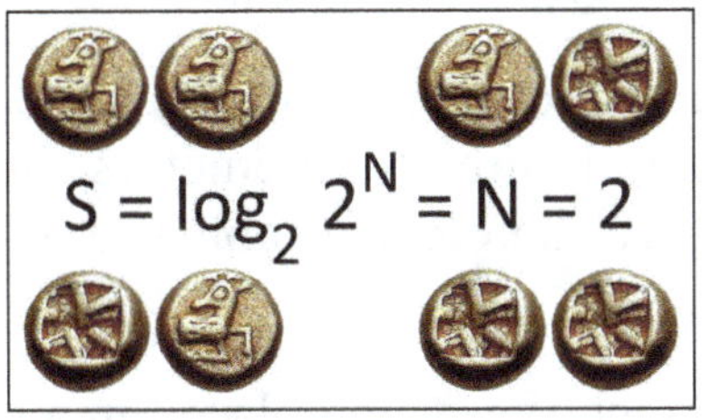

2 bits of information entropy (if coins are fair).

The third equation, originally formulated by Ludwig Boltzmann between 1872 and 1875, is so iconic that it appears on Boltzmann's tombstone. W instead of the modern Ω for what is often called the 'number of available states') The existence of two definitions (k ln Ω and ∫dQ/T) for the same entity leaves one wondering if they are actually the same thing. Adding to the complexity is the fact that Ω in equation $S = k_B \ln \Omega$ has a number of definitions. For example, in information theory the constant, k_B, equals one, and entropy is measured in 'bits' if the base of the logarithm is 2 (instead of base-*e* as used in physics). Information Ω is the inverse of probability if all probabilities are equal. And if all probabilities are equal, then Ω is the number of possible outcomes. The number of fair coin tosses is its entropy in bits. (Since there are 2^N outcomes for *N* fair coins, and $log_2 N^2 = N$). In physics, entropy can be troublesome to calculate primarily due to issues concerning the appropriate number of equally probable states. Another vexing problem is the fact that the equations of Newtonian and Hamiltonian physics possess time reversal symmetry, while reality apparently does not. Entropy reflects that reality by only increasing as time evolves.

Thermodynamic Definition of Entropy

State variables: Pressure (P), Energy (E), Volume (V), and Temperature (T).

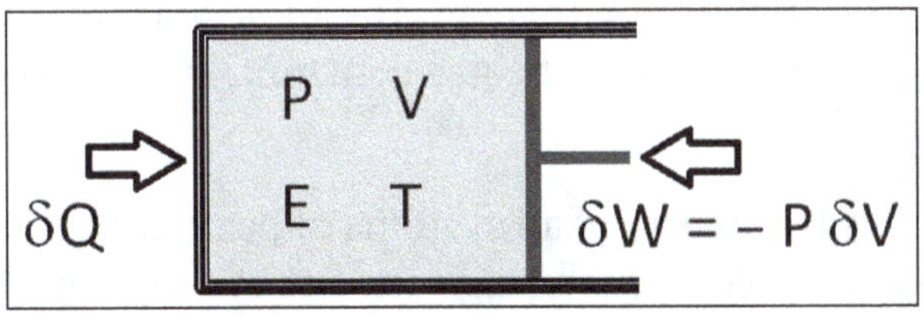

Pressure, Energy, Volume, and Temperature are state variables. Work (W) and heat (Q) are not.

A state variable is a measurable physical property that can be uniquely defined for a substance that is in thermal, mechanical, and chemical equilibrium. We shall refer to this substance as ‹the system› and two systems play a key role in this discussion:

- The monatomic ideal gas. The ideal gas model tends to fail at lower temperatures or higher pressures, when intermolecular forces and molecular size become important, and more complex ideal gasses, such as the diatomic molecule require quantum mechanics to understand rotational energy. These complications render such systems unsuitable to be fully modeled.
- At the opposite extreme, we are investigating any system (however complex) in what is

called thermodynamic equilibrium. One requirement for being in thermodynamic equilibrium is that the pressure and temperature are both uniform throughout the system. Thermodynamics is often applied as an approximation when these conditions do not hold (for example when calculating the speed of sound).

Pressure and volume can be measured by observing the force and location of the piston, respectively. Temperature was originally measured with thermometers, but for our purposes the ideal gas thermometer is more convenient. We therefore establish the following operational definition of temperature:

The temperature of a substance is such that when placed in a thermal contact with a monatomic heat bath, $dQ = 0$ where dQ is heat flow, and temperature is found by solving the equation of state for an ideal monatomic gas:

$PV = nRT = Nk_BT$ (Equation of state for ideal gas).

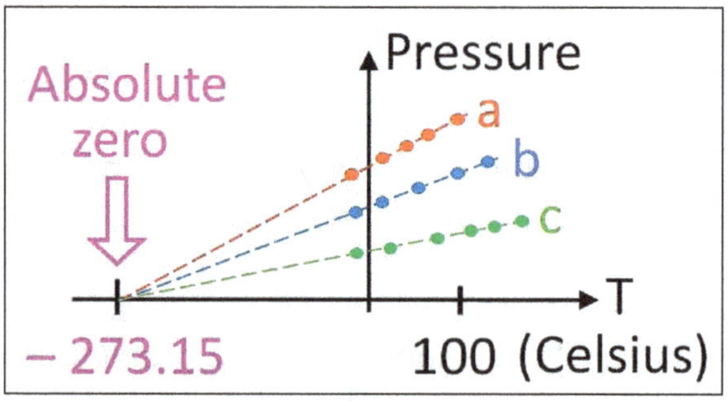

Plots of pressure vs. temperature for three different gas samples at the same volume extrapolate to absolute zero.

Here, N is the number of particles (assumed constant in this introduction to classical thermodynamics), n is the number of moles, and R is called the gas constant. When classical thermodynamics was being constructed in the eighteenth and nineteenth centuries, it was not yet known that one mole consists of 6.022 × 1023 atoms. Instead, the mole was defined as equal to the number of carbon atoms in 12 grams of carbon, and the gas constant, R, was an experimentally known parameter equal to 8.314 $JK^{-1}mol^{-1}$.

We shall assume that energy is a property of all systems of particles and that energy can be uniquely determined by other state variables. This brings our list of state variables to four:

- Volume (*V*)
- Pressure (*P*)
- Temperature (*T*)
- Energy(*E*)

Any function of these state variables is also a state variable. Generally speaking, any two state variables is sufficient to determine the system. One exception is energy for an ideal gas, where since energy is directly proportional to temperature, energy and temperature are insufficient for determining the state of an ideal gas.

Heat, Work and the Heat Engine

Heat, dQ, is the flow of energy from a hot to a cold substance. (The fact that heat never flows between objects with the same temperature is part of the zeroth law of thermodynamics.) Work is $dW = Fdx = -PdV$, where the minus sign ensures that dE, dW, and dQ are all positive if they add energy to the system. Since work and heat are the only two ways energy can enter or leave this sealed system, we have,

$dE = dQ + dW$ First law with differentials; eventually becomes $dE = TdS - PdV$.

Why Heat and Work are not State Variables

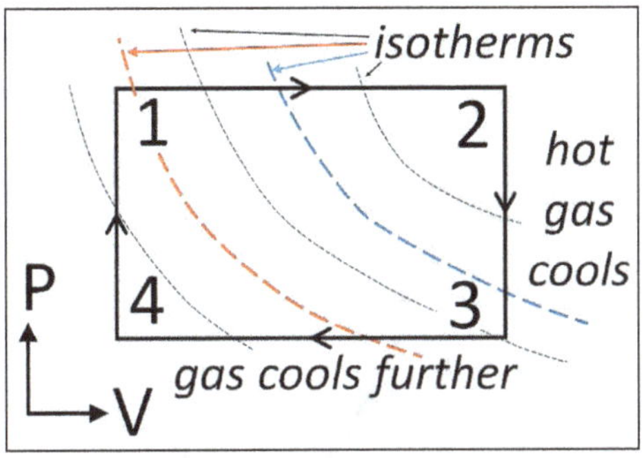

In the 1-2-3-4 direction, this serves as a heat engine, Isotherms indicate curves of equal temperature.

This figure shows pressure (P) as positive, but an equivalent P-V diagram could be constructed for an substance with negative pressure, such as a rubber band. The rubber band is a good example of approximate thermodynamic equilibrium because the system is never properly in chemical equilibrium - the cycles would not exactly close upon themselves as the rubber degrades into a more stable (but less desirable) substance. Nevertheless, one can construct a rubber band heat engine or heat pump: Stretch a thick rubberband and hold it stretched until it cools off to room temperature. Then decompress the rubberband and press it to your lips. It will feel cool because you have created a heat engine in reverse, called a heat pump *(or a refrigeration unit).*

To understand why heat, ΔQ, and work ΔW are not state variables,we introduce the heat engine, which is a cycle that converts heat into work (or it can act as a heat pump if the cycle is reversed.) Since only two state variables are independent, it is possible to fully specify a system as a point on a two dimensional graph, for example on the P-V (pressure-volume) diagram shown in the figure. The other state variable (T) can be depicted using contours on the graph. Contours of constant temperature are called isotherms. For the ideal gas, these contours are hyperbolas, but we are not restricting our discussion to the ideal gas.

The cycle shown in the figure to the right is an Ideal Heat Engine. It is 'ideal' only in the sense that certain aspects are easy for students to calculate. From points 1-2, energy in the form of work leaves the system. No work is done from points 2-3 because the piston remains motionless. Work energy does re-enter the system from points 3-4, but less work is involved because the pressure is lower.

The work done by the engine per cycle equals to the area enclosed by the cycle on a P-V diagram.

This equivalence of work performed to area inside the cycle on a P-V plot is a general property of heat engines.

By conservation of energy, the work performed by the engine must equal the energy that enters, and this energy can only be delivered into the system as heat. Since work energy leaves the engine at each cycle, and heat energy enters, work and heat are not state variables. It is meaningless to ask how much 'heat is in a system'. It is best to always write work and heat using notation that reflects this fact, e.g., ΔW and ΔQ.

How do we know that entropy even exists? The Clausius definition of entropy as the integral, $dS = \int dQ/T$, defines entropy only as a differential. As we have seen, this does not prove that entropy is a state variable.

Carnot's Theorem

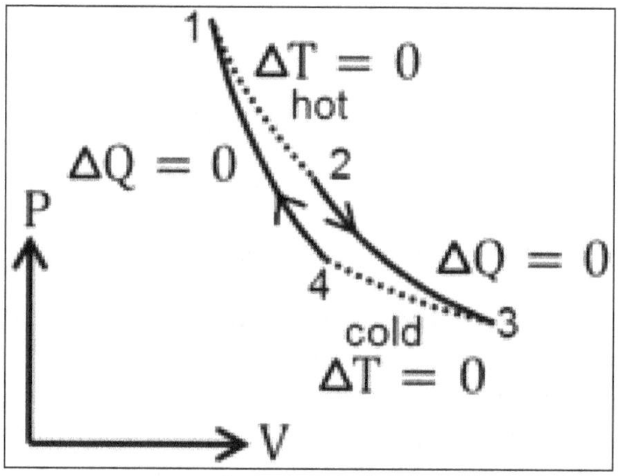

The Carnot cycle.

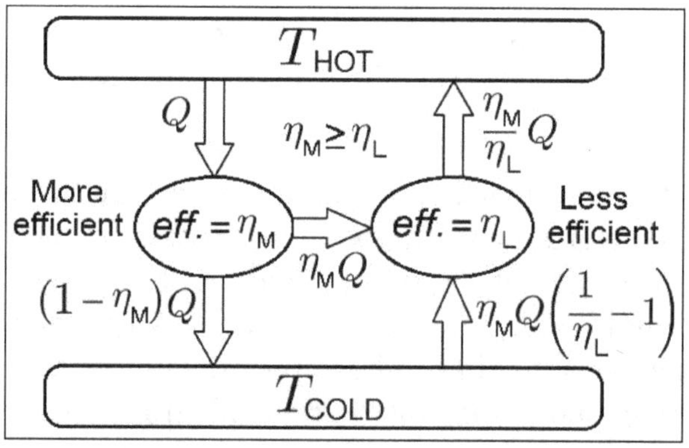

Two irreversible heat engines with different efficiency will violate the second law of thermodynamics when the high efficiency engine (H) drives the low efficiency device (L) as an air conditioner.

The 'ideal heat cycle' described above is easy to understand because work is so easy to calculate. Another heat engine is far more important. The Carnot cycle consists of four legs. Two legs are 'isotherms' with $\Delta T=0$, (temperature remains constant), and two legs are 'adiabats' with $\Delta Q=0$ (no heat flows). It can be shown that the Carnot cycle is the most efficient way to do work by transferring heat from a reservoir at T_{hot} to a reservoir T_{cold}. But a far more important property of the Carnot cycle is the fact that all substances have exactly the same efficiency when used in a Carnot cycle. This can be proven using the diagram shown. If two substances produced work with different efficiency, one could use the less efficient substance in a heat pump (i.e., air conditioner) and violate the second law of thermodynamics generating useful work from a temperature reservoir without the requirement that heat must be dumped into a colder reservoir.

The efficiency of a heat engine operating between two temperatures is the ratio net work (per cycle) to the heat flow out of the hot reservoir (per cycle). Ideally, one would want to convert all the heat from the hot reservoir into work and not 'dump' any 'waste heat' into the cold reservoir.

Why the Carnot Engine is Reversible?

Key to the proof is the fact that the second engine in the figure to the right is being operated in reverse as a [w: heat pump|]. Carnot's theorem does not hold if the engine is not reversible. The Carnot cycle shown in the figure to the left is reversible because heat enters at constant temperature. Strictly speaking, the Carnot cycle is only reversible if the temperature of the substance equals the temperature of the heat reservoir. It may seem odd that two systems (reservoir and substance inside the piston) are exchanging heat when both are at the same temperature, as heat only flows between bodies of unequal temperature. This paradox is resolved by admitting that the two temperatures are not *exactly* equal, but only *approximately* equal. And since a heat flow is often directly proportional to the difference in temperature, a true Carnot cycle would almost take an infinite amount of time to accomplish. The Carnot engine is truly an idealized and theoretical construct that can never truly exist. But something arbitrarily close to a Carnot engine can be built, and will operate in reverse in approximately the same fashion.

Why the Carnot Engine is Important?

It is not uncommon for a heat engine to employ a complicated substance (e.g. Freon). Carnot's theorem is a statement about all substances that remain in equilibrium throughout the cycle of a heat engine. If we can fully understand just one simple substance, then we can learn something that is true for all substances in equilibrium. The thing that we learn about will be called entropy. And the substance that we shall (attempt to) fully understand is the monatomic ideal gas.

Carnot Cycle

Carnot described the reversible cycle undergone by a fluid, with seven states, but equivalent to what is known today as the Carnot cycle. Two bodies were needed: body A (the hot source) at temperature T_A, and body B (the cold source) at temperature $T_B < T_A$. The original description of the cycle is intricate, but Carnot noted that such a cycle could be simplified if $T_A \approx T_B$. In all his calculations, he considered a cycle with the following stages:

- Contact between the gas abcd and the body A; movement of the piston from cd to ef.
- Removal of the body A; contact between the gas in abef with the body B; return of the piston from ef to cd.
- Removal of the body B; contact between the gas and the body A; movement of the piston from cd to ef. In other words, the first stage is repeated, and so on.

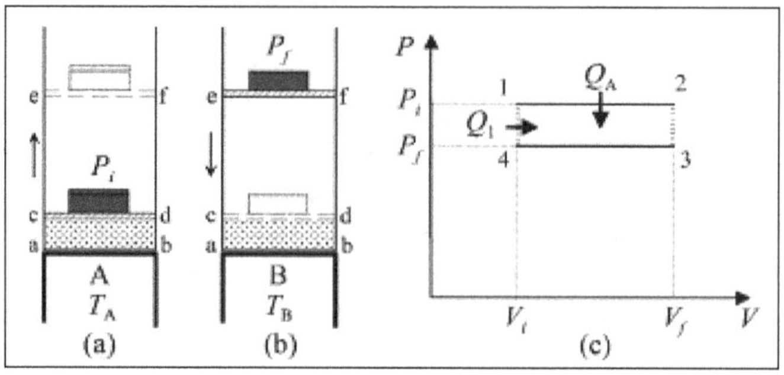

Cycle described in Carnot's memoir (modified Carnot cycle). (a) Isobaric expansion in diathermal contact with hot reservoir A at temperature T_A; (b) isobaric compression in contact with cold reservoir B at temperature T_B; (c) Clapeyron's diagram for the cycle.

Figure shows the modified Carnot cycle in a (P, V) diagram. The process $1 \rightarrow 2$ is a quasi-static isobaric expansion at pressure Pi in contact with the hot reservoir; in $2 \rightarrow 3$, the temperature decreases at constant volume V_f !irreversible isochoric"; the process $3 \rightarrow 4$ is a quasi-static isobaric compression at a lower pressure P_f in contact with the cold reservoir until the initial volume V_i is recovered; finally, in $4 \rightarrow 1$, the temperature increases at constant volume "irreversible isochoric" until the initial temperature is reached.

Carnot considered cycles for air, water, and alcohol with $T_A - T_B = 1$ °C, which is rather small in comparison with T_A and T_B. As we shall see, replacing the Carnot cycle by the modified Carnot cycle is quite reasonable for water and alcohol, where liquid–vapor equilibrium phase transitions are present. On the other hand, the modified cycle is impossible for air if it is in contact with bodies A and B only.

Carnot implicitly defined the efficiency of the cycle as the ratio of the net !or useful" work delivered by the cycle to the heat taken from the hot reservoir:

$$\eta = \frac{W_N}{Q_1 + Q_A}.$$

The substance absorbs heat in one of the isochoric processes (Q_1) and in the expansion in contact with body A (Q_A), but in many cases, Carnot assumed that Q_1 could be neglected in comparison with Q_A.

For W_N he took the difference between the work in processes $1 \rightarrow 2$ and $3 \rightarrow 4$. He knew that the work could be calculated from $W = P_{ext} \Delta V$, where P_{ext} is the external pressure and ΔV the volume variation. Hence,

$$W_N = \Delta P \Delta V,$$

where, $\Delta P = P_i = P_f$ and $\Delta V = V_f - V_i$. This work is given by the area 1234 in figure.

Carnot's Calculations

Carnot studied cycles with air between 0 and 1 °C, with water (liquid-steam) in three cycles between 99 and 100 °C, between 0 and 1 °C, and between 77 and 78 °C, and with alcohol (liquid–vapor) between 77.7 and 78.7 °C.

Cycle with Air between 0 and 1 °C

Carnot considered 1 kg of air at 0 °C and volume V_i (0 °C) = 77 m³, which is heated at constant volume, until it reaches 1 °C, absorbing heat Q_1 (which he neglected for the efficiency). Then, at constant pressure (P_i = 1 atm = 760 mm Hg = 1.013 × 10⁵ Pa), C A absorbing heat Q_A. Carnot assumed that this expansion involved two contributions, one isothermal, with ΔV = V_i/267, and another, adiabatic, with ΔV = V_i/116, estimating the total volume increase as:

$$\frac{\Delta V}{V}=\frac{1}{267}+\frac{1}{116}=1.25\times10^{-2}\Rightarrow\Delta V=9.52\times10^{-3}\,\mathrm{m}^3.$$

To calculate ΔP, Carnot used Gay-Lussac's rule. At constant volume, the relative pressure increment corresponding to the temperature variation from 0 to 1 °C was taken to be ΔP/P = 1/267. Hence,

$$\Delta P=\frac{1}{267}10.4=3.89\times10^{-2}\,\mathrm{m.w.}=3.79\times10^{2}\ \mathrm{pa}.$$

[As pressure unit Carnot used the meter of water (m.w.): 1 atm = 10.4 m.w.] Thus, from equation $W_N=\Delta P\Delta V$, Carnot calculated the useful work done in cycle 1234:

$$W_N^{(1)}=3.72\times10^{-4}\ \mathrm{m}^3\,\mathrm{m.w.}=3.61\,\mathrm{J}.$$

To evaluate the absorbed heat, Carnot assumed that the heat capacity of air was 0.267 times the heat capacity of an equal mass of water. Taking c_{air} = 0.267 kcal kg⁻¹ °C⁻¹, his estimate for the heat provided by the hot body was,

$$Q_A^{(1)}=mc_{air}\Delta T=0.267\ \mathrm{kcal}=1116.1\ \mathrm{J}.$$

The work ($dU = TdS - pdV + \mu dN + vdp + \omega dL + \mathbb{E}dP + HdM$) was correctly determined but not the heat ($Q_A^{(1)}=mc_{air}\Delta T=0.267\ \mathrm{kcal}=1116.1\ \mathrm{J}$). In Appendix B we compute the work and the heat absorbed in the cycle and compare them with Carnot's results.

From ($W_N^{(1)}=3.72\times10^{-4}\ \mathrm{m}^3\,\mathrm{m.w.}=3.61\,\mathrm{J}$) and ($Q_A^{(1)}=mc_{air}\Delta T=0.267\ \mathrm{kcal}=1116.1\ \mathrm{J}$), Carnot obtained the efficiency of the cycle:

$$\eta^{(1)}=\frac{W_N^{(1)}}{Q_A^{(1)}}=3.25\times10^{-3}.$$

(Although efficiencies are in fact dimensionless, those reported by Carnot were expressed in m³ m.w. kcal⁻¹. Because 1 m³ m.w. = 9800 J and 1 kcal = 4180 J, Carnot's efficiencies should be multiplied by 2.34 to become pure numbers.)

Cycle with Water between 99 and 100 °C

The second cycle used 1 kg of water. The reservoir temperatures were t_A = 100 °C and t_B = 99 °C. During the cycle, water undergoes liquid–vapor phase transitions. In order to evaluate the useful work, Carnot had to calculate both the volume increment (from liquid to steam) and the pressure increment (from steam pressure at 99 °C to steam pressure at 100 °C).

For steam, he assumed the ideal gas equation. The specific volume of steam at 100 °C and atmospheric pressure is $\upsilon_V(100\ °C) = 170\ m^3\ kg^{-1}$. Taking the specific volume of liquid water $\upsilon_L(99\ °C) = 1 \times 10^{-3}\ m^3\ kg^{-1}$, he estimated the volume increment in the cycle to be $\Delta V \approx V_V(100\ °C) = 170\ m^3$. To obtain the pressure increment, Carnot used liquid–vapor equilibrium data from Biot. He fitted the temperature dependence of the steam volume, υ_V, by the function.

$$\upsilon_V = 10^{a(t)},\ a(t) = \frac{2268 - 1964t}{1000 + 3.30t},$$

where υ_V is in liters per gram and t in degrees Celsius. To calculate the steam pressure at 99 °C, Carnot first inserted this temperature in equation above and then introduced the resulting specific volume, $\upsilon_V(99\ °C) = 1.754\ l\ g^{-1} = 1.754\ m^3\ kg^{-1}$, together with the temperature, in the equation of state.

$$P_V(t) = \frac{3.52}{\tilde{o}_V}(267 + t),$$

obtaining $P_V(99\ °C) \approx 734$ mm Hg. Because, on the other hand, $P_V(100\ °C) = 760$ mm Hg, he got the pressure increase: $\Delta P = 26$ mm Hg $= 3.465 \times 10^3$ Pa.

The net work produced in the cycle by 1 kg of liquid water at 99 °C, which transforms into steam at 100 °C, was $W_N^{(2)} = 1.70 \times 0.36 = 5.890 \times 10^3$ J.

Moreover, Carnot knew that the heat required to transform into steam 1 g of liquid water at 100 °C was 550 kcal kg⁻¹ (specific enthalpy of vaporization in table). Thus, $Q_A^{(2)} = 2299.0 \times 10^3$ J. The cycle efficiency was,

$$\eta^{(2)} = \frac{W_N^{(2)}}{Q_A^{(2)}} = 2.57 \times 10^{-3}.$$

Cycle with Water between 0 and 1 °C

Carnot next considered a cycle with liquid water in equilibrium with its steam, between 0 and 1 °C. He compared this efficiency with that of the first cycle with air between the same temperatures.

Taking t=1 °C in Equation $\upsilon_V = 10^{a(t)},\ a(t) = \frac{2268 - 1964t}{1000 + 3.30t}$, one obtains the steam specific volume $\upsilon_V(1\ °C) = 174.16\ l\ g^{-1} = 174.16\ m^3\ kg^{-1}$. Inserting this value and t = 1 °C in Equation $P_V(t) = \frac{3.52}{\upsilon_V}(267 + t)$, one has $P_V(1\ °C) = 5.418$ mm Hg. On the other hand, because from Biot's data $P_V(0\ °C) = 5.06$ mm Hg, one finds $\Delta P = 0.358$ mm Hg $= 47.71$ Pa.

Given $\upsilon_L(0\ °C) \approx 1 \times 10^{-3}\ m^3\ kg^{-1}$, one has $\Delta V \approx V_V(1\ °C) = 174.16\ m^3$. Introducing these pressure and volume increments in Equation $W_N = \Delta P \Delta V$, the net work is $W_N^{(3)} = 8.310 \times 10^3$ J. Carnot did not explicitly present these calculations but obtained a value that is similar after convenient translation.

He assumed that the heat absorbed in the cycle was the sum of two contributions: One part would increase the water temperature from 0 to 100 °C at normal pressure, that is, 100 kcal kg^{-1}; and the other would evaporate water at 100 °C, that is, 550 kcal kg^{-1}. At his time, it was believed that the heat required to vaporize water was independent of the pressure: to vaporize water at 1 °C, one would need the same heat that would be required first to raise the water temperature, up to the normal boiling point, and then to vaporize it completely, always at normal pressure. The total heat absorbed in the cycle should be 650 kcal kg^{-1} = 2717 kJ kg^{-1}.

Finally, for this water cycle, the efficiency was $\eta = \frac{W_N}{Q_1 + Q_A}$. (Carnot arrived at a slightly different value),

$$\eta^{(3)} = \frac{W_N^{(3)}}{Q_A^{(3)}} = 3.06 \times 10^{-3}.$$

Cycle with Alcohol between 77.7 and 78.7 °C

The next system studied by Carnot was 1 kg of ethylic alcohol. For alcohol, whose normal boiling point is 78.7 °C, he assumed that the vapor pressure at 77.7 °C was reduced by 760/25 mm Hg: ΔP = 30.4 mm Hg = 4.05 × 10^3 Pa.

To obtain the volume variation when alcohol goes from liquid to vapor, one may use Equation $P_V(t) = \frac{3.52}{\upsilon_V}(267 + t)$, adapted for alcohol. The overall numerical factor of that equation should be multiplied by the ratio of water and alcohol molecular weights:

$$\upsilon(P,t) = 1.37 \frac{267 + t}{P}.$$

Actually, this equation was not used by Carnot. According to Equation above, the volume occupied by alcohol vapor near its boiling point, t = 78.7 °C, is υ_V(78.7 °C) = 0.626 m^3 kg^{-1}. Thus, $\Delta V \approx V_V$ (78.7 °C) = 0.626 m^3, because υ_L(77.7 °C) ≈ 0.74 × 10^{-3} m^3 kg^{-1}. The net work in this cycle was $W_N^{(4)}$ = 2.54 × 10^3 J. Again, this work was not explicitly given by Carnot.

On the other hand, Carnot knew that the heat required to evaporate 1 kg of alcohol at its boiling point was $Q_A^{(4)}$ = 207 kcal = 865.3 × 10^3 J. The efficiency of the cycle was then,

$$\eta(4) = \frac{W_N^{(4)}}{Q_A^{(4)}} = 2.94 \times 10^{-3}.$$

Carnot again presented a slightly different value.

Cycle with Water between 77 and 78 °C

To compare his results for alcohol and water, Carnot studied a fifth cycle with liquid water at 77 °C, which evaporates at 78 °C. First, he sought the specific volumes occupied by steam at those temperatures. Using Equation $\upsilon_V = 10^{a(t)}$, $a(t) = \frac{2268 - 1964t}{1000 + 3.30t}$, he found the specific volumes υ_V(77 °C)

= 4.005 1 g^{-1} and υ_V(78 °C) = 3.849 1 g^{-1}, respectively. Using Equation $P_V(t) = \frac{3.52}{\upsilon_V}(267 + t)$, he could obtain the corresponding vapor pressures for water steam at those temperatures: P_V(77 °C) = 302.3 mm Hg and P_V(78 °C) = 315.5 mm Hg, respectively. Thus $\Delta V \approx V_V$(78 °C) = 3.849 m^3, and the pressure difference for water steam between 77 and 78 °C is ΔP = 13.2 mm Hg = 1.76×10^3 Pa. So the useful work performed in this cycle was $W_N^{(5)} = 6.77 \times 10^3$ J.

Carnot obtained $Q_A^{(5)}$ = 572 kcal = 2390×10^3 J. Thus, the efficiency of this cycle was,

$$\eta^{(5)} = \frac{W_N^{(5)}}{Q_A^{(5)}} = 2.83 \times 10^{-3}.$$

These calculations were not presented by Carnot. He just gave the cycle efficiency.

Motion in Phase Space

Both the derivation of the Maxwell-Boltzmann equation for energy distribution, as well as the calculation of entropy in would today be called the spirit of information entropy require a deep understanding of how particles move in phase space. Phase space typically has twice the number of dimensions as the 'configuration space' students initially use to visualize motion. For example, instead of viewing motion in the two dimensional (x-y) plane, phase space would include four dimensions (x, y, v_x, v_y).

Ergodic Hypothesis and Second Law of Thermodynamics

Why can't we trap atoms the way we trap fruit flies?

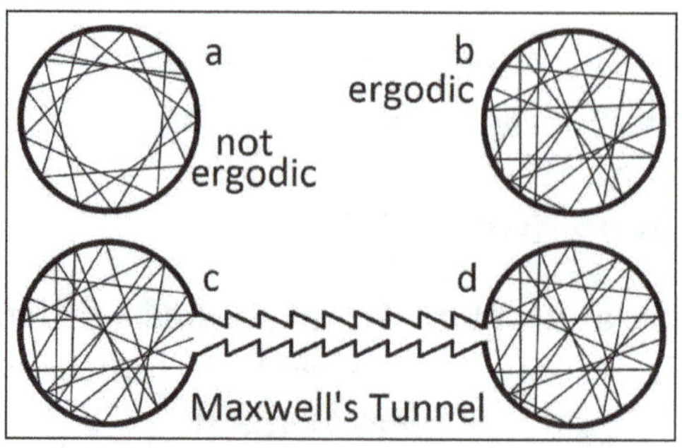

Ergodic hypothesis w reflecting rays.

For simplicity we develop ideas with unphysical models. Consider a two dimensional gas of rays for which speed is irrelevant. They never collide, and change direction only after hitting a wall. If the walls are perfectly smooth, most particles would follow paths resembling that shown in (a) of the figure. Such an orbit does not occupy all of its available phase space, and is therefore called *non-ergodic*. Not only do these paths fail to fill the available phase space, they do not seem uniformly fill the space available to them. But if the smooth surfaces have any flaws, no matter how small, a gas of such particles would eventually fill the area uniformly, as shown in (b). In classical thermodynamics, the assumption is made that particles fill their available phase space uniformly in this fashion. This is called the ergodic hypothesis, and it was originally proposed by L. Boltzmann.

Although we do not yet know why, it is clear that any *trick* to coax the gas from one of the two connected circles (c) and (d) using an artfully shapted *tunnel* as shown in the figure would fail. If such a tunnel ever managed to create a significant pressure difference, it could drive a perpetual motion machine of the second kind. An automobile›s pistons would be driven by arranging for the atoms to randomly find their way to one side of a piston. The compressed air would expand to power the car. In expanding in this fashion the air would cool. But the exhaust from this engine would be cold air, some of which could be pumped directly into the cab on a hot summer day. The only fuel required is the random motion of atoms that occurs everywhere on earth. The impossibility of ever creating such a device shall be our working definition of the second law of thermodynamics.

The term 'ergodic hypothesis' is somewhat of a misnomer because the word 'hypothesis' often suggests either a controversy or unresolved scientific question. Often it is. For our purposes, it is best to view the 'ergodic hypothesis' as an assumption that may or may not hold for a given system.

The ergodic hypothesis can be stated as: All regions of phase space that are accessible to a given particle are equally populated, provided that phase space is expressed in Hamiltonian canonical variables.

Canonical Variables, Liouville's Theorem and the Ergodic Hypothesis

This equality in occupation of phase space is a consequence of Liouville's theorem, and requires that the canonical variables of a Hamiltonian dynamics be used. Liouville's theorem does not prove the Ergodic hypothesis.

Moreover, it likely that no proof of the ergodic hypothesis exists, since it seems to violated for certain Hamiltonians, such as that used in the Fermi-Pasta-Ulam problem. But Liouville's theorem does ensure that once the probability becomes uniform throughout a region of phase space that is accessible to a particle, the ergodic hypothesis will continue to hold.

We shall assume the ergodic hypothesis throughout this introductory look at Statistical thermodynamics. And since this is an introduction to the subject, we shall not prove whether or not the variables are such that Liouville's theorem applies. For example, position and velocity (x,v) are not proper canonical variables, but v is directly proportional to p=mv, phase space in (x,v) obeys Liouville's theorem and we there assume that the ergodic hypothesis applies. Kinetic energy is not a suitable variable for a phase space where the ergodic hypothesis can be applied.

Probability Distribution Functions

The probability of having a certain velocity in a one dimensional gas is described by a distribution function. Letting Prob denote probability, we have:

$$\text{Prob}(\upsilon_1 < \upsilon_x < \upsilon_2) = \int_{\upsilon_1}^{\upsilon_2} f_\upsilon(\upsilon)dv \to f_\upsilon(\upsilon)\Delta\upsilon$$

The 'rightarrow' $(\to)$ holds when the range of velocity is so small that $f_\upsilon(\upsilon)$ is nearly uniform. The subscript on the probability distribution function, f_υ will permit us to simultaneously talk about probability distributions for different variables.

Two Dimensional Gas Confined to within a Circle

The probability density for one particle confined to a circle of radius R can be found by satisfying two criteria:

- The integral over all accessible regions of xy space is one.
- The system is ergodic if proper variables are used; we take x and y to be proper variables without proof.

Hence,

$$f_{xy}(x,y)dx\,dy = \frac{1}{\pi R^2}dx\,dy$$

It is useful to change variables using $x = r\cos\theta$ and $y = r\sin\theta$,

$$dx\,dy = \det\begin{bmatrix} \frac{\partial x}{\partial r} & \frac{\partial x}{\partial \theta} \\ \frac{\partial y}{\partial r} & \frac{\partial y}{\partial \theta} \end{bmatrix} dr d\theta = \det\begin{bmatrix} \cos\theta & -r\sin\theta \\ \sin\theta & r\cos\theta \end{bmatrix} dr d\theta = r dr d\theta$$

Hence,

$$f_{r\theta}(r,\theta)drd\theta = \frac{r}{\pi R^2}drd\theta$$

Note that $r\theta$ space is not uniformly occupied.

Kinetic Energy Probability Distribution

Letting $E = \frac{1}{2}mv^2$ equal kinetic energy, we solve for $v = (2E/m)^{1/2}$ the probability distribution in energy can be calculated as follows:

$$f_\epsilon(\epsilon)d\epsilon = 2f_\upsilon(\upsilon)dv.$$

The peculiar factor of 2 occurs because two values of v contribute to the same kinetic energy. Since this is a one dimensional calculation, it is proper refer to ϵ as kinetic energy for one degree of freedom. (In this way, a free particle in three dimensions has three degrees of freedom.) Using, $d\epsilon = mvdv$, and a bit of algebra, it can be shown that:

$$f_\epsilon(\epsilon) = \sqrt{\frac{2}{m\epsilon}} f_\upsilon(\upsilon).$$

Maxwell's Symmetry Argument for Velocity Space

Maxwell's original derivation of the Maxwell-Boltzmann equation relied on the concept that the product of two or more independent events occurring equals the product of the probability of each event. For example the probability of flipping a fair coin twice and obtaining two heads (H-H) is the product of obtain a heads one coin squared: P(H-H)=P(H)P(H) = ½·½=¼. He reasoned that for N particles, the probabilities distributions for all the particles were independent, and therefore obeyed a similar product rule:

$$F_{\{\upsilon\}}(\upsilon_1, \upsilon_2, \ldots \upsilon_N) = \prod_{n=1}^{N} f_\upsilon(\upsilon_n) = f_\upsilon(\upsilon_1) \cdot f_\upsilon(\upsilon_2) \cdot \ldots \cdot f_\upsilon(\upsilon_N)$$

where,

$$\{\upsilon\} = [\upsilon_1, \upsilon_2, \ldots \upsilon_N]$$

represents one point in a very large number of dimensions. The use of a capital letter for this 'grand' distribution function reflects the relative complexity of this space. Boltzmann also argued that the 'grand' distribution function depended only on the total energy of the gas:

$$F_{\{\upsilon\}} = (\upsilon_1, \upsilon_2, \ldots \upsilon_N) = F_{\{\upsilon\}}(E),$$

where,

$$E = \frac{1}{2}mv_1^2 + \frac{1}{2}mv_2^2 + \ldots + \frac{1}{2}mv_N^2 = \epsilon_1 + \epsilon_2 + \ldots + \epsilon_N$$

is the total energy. Ordinarily the probability distribution would also depend on density, but an ideal gas is assumed to be sufficiently tenuous that collisions have no impact except that they eventually establish thermal equilibrium. It is the total energy of an ideal gas of N particles that uniquely determines how hot it is.

Carefully Redefining Functions

The micro distribution functions, $f(\upsilon_n)$, also depend only on speed, and hence only on the energy of each atom, $\epsilon_n = mv_n^2/(2m)$. But extreme care must be used in expressing this fact. It is safest to define a new function, g, as follows:

$$g(\epsilon) = f_{\tilde{o}}\left(\sqrt{\frac{2\epsilon}{m}}\right).$$

This function, g, is numerically equal to the micro distribution for a single particle, but as we shall see, is not the distribution function for the kinetic energy of an atom. Also, to be precise, it is necessary to think of F as one of two functions:

$$F = F_{\{\epsilon\}} = F(\epsilon_1, \epsilon_2, ... \epsilon_N) = g(\epsilon_1) \cdot g(\epsilon_2) \cdot ... \cdot g(\epsilon_N).$$

$$F = F_E = F(E) = F(\epsilon_1 + \epsilon_2 + ... + \epsilon_N).$$

Velocity Distribution Function for a 1-D Ideal Gas

Using the first form, we evaluate the variation of F with ϵ_1 as:

$$\frac{\partial F}{\partial \epsilon_1} = \frac{dg}{d\epsilon_1} \cdot g(\epsilon_2) \cdot .... \cdot g(\epsilon_N) = \frac{1}{g_1} \frac{dg_1}{d\epsilon_1} F.$$

Using the second form,

$$\frac{\partial F}{\partial \epsilon_1} = \frac{dF}{dE} \frac{dE}{d\epsilon_1} = \frac{dF}{dE}.$$

Similarly we can prove that $\partial F / \partial \epsilon_2 = dF / dE = \partial F / \partial \epsilon_1$, which means that two expressions, each (exclusively) involving different variables are equal:

Where $f_\upsilon(\upsilon)$ is evaluated at $\upsilon = 2\epsilon / m)^{1/2}$.

$$\frac{1}{g_1} \frac{dg_1}{d\epsilon_1} = \frac{1}{g_2} \frac{dg_2}{d\epsilon_2} = -\kappa$$

where we have yet to establish that the constant, κ is a positive number. But solving this differential equation, we have:

$$g(\epsilon) = Ae^{-\kappa\epsilon}$$

Using the Ideal Gas Law to find Kappa

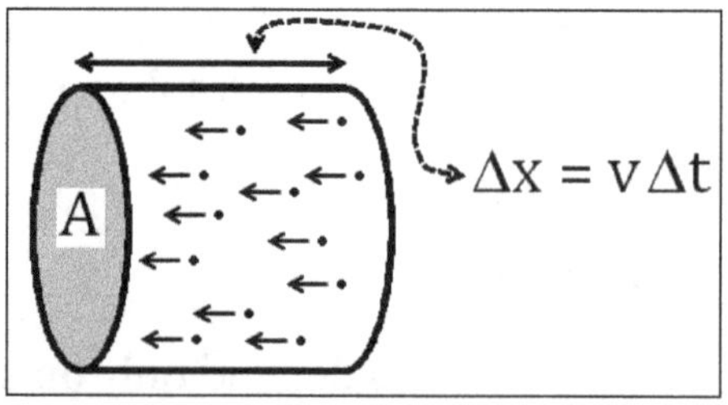

A force is exerted as these particles bounce off the wall. For a single particle, the momentum transferred is 2mv/Δt, where Δt=2L/v is the time between collisions if the gas is contained by a cube of length, L. The pressure on one wall is is therefore mv_x^2/L^3.

The ideal gas law is an observable fact about actual gasses. To establish that $\kappa = m/(2k_B T)$ we must perform a theoretical calculation of the pressure using this distribution function. We use the fact that the distribution functions are now known (with κ as a parameter) and calculate the pressure exerted by this gas by employing the momentum transfer associated with elastic collisions of particles of a known density and velocity distribution off a wall, as shown in the figure.

Hence we have the probability distribution function for the velocity of a one-dimensional ideal gas expressed in terms of temperature:

$$f_\upsilon(\upsilon) = A\exp(-\epsilon / k_B T),$$

where, $A = \left(\frac{1}{2\pi m k_B T}\right)^{1/2}$, is a normalization constant that can be found by setting $\int_{-\infty}^{\infty} f_{\bar{o}} dv = 1$.

The Importance of using Proper Phase Space Variables

If we change variables using the methods described above, we establish that,

$$f_\epsilon(\epsilon) \neq A\exp(-\epsilon / k_B T)$$

In other words, the wrong answer would have resulted if we assumed if this calculation had begun with slightly different (but incorrect) premise: $f_\epsilon(\epsilon) = f_1(\epsilon_1) f_2(\epsilon_2) \ldots f_N(\epsilon_N)$. This premise is false because the transformation from the permissible phase space variables $(\upsilon_1, \ldots \upsilon_N)$ to energy variables introduces the factor $\epsilon^{-1/2}$.

The Velocity Distribution if the Gas is Confined by a Potential

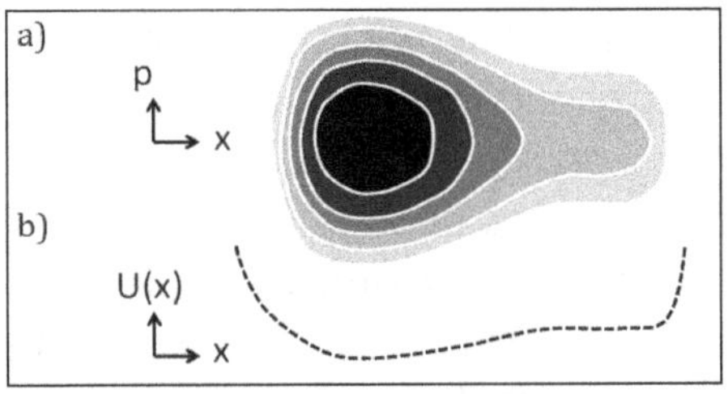

Maxwell Boltzmann distribution for one particle confined by a generic potential. The system is ergodic because particles follow contours of constant density in phase space.

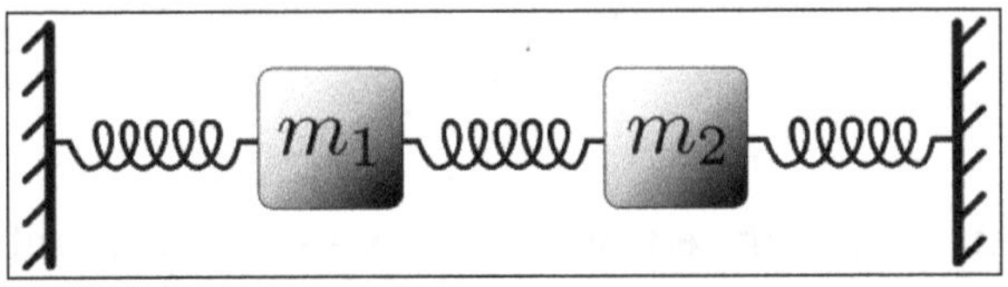

A canonical transformation can convert these two coupled oscillators into two uncoupled oscillators with different frequencies of oscillation.

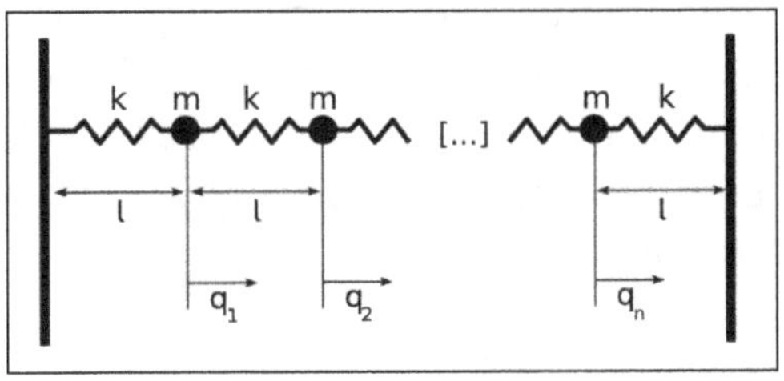

Kontinuumlimes

If this calculation is repeated for a one dimensional gas confined by a potential well, $\Phi = \Phi(x)$, one obtains the following distribution function:

$$f_\upsilon(\upsilon) = \text{Constant} \cdot \exp\left(-\frac{\frac{1}{2}mv^2 + \Phi(x)}{k_B T}\right).$$

This is a significant result because the potential can be that of a harmonic oscillator $\left(\Phi = \frac{1}{2}k_S x^2\right)$ where k_S is the spring constant. Moreover, a Hamiltonian exists that describes two coupled oscillators as two uncoupled oscillators. And, this method can be extended to include a very large number of coupled oscillators. In other words, a linear wave can be viewed as a classical gas of uncoupled harmonic oscillators. It was Plank's attempt to perform an thermodynamic analysis of light waves in thermal equilibrium with a closed box that lead to the breakdown of classical physics.

Thermodynamic Equilibrium

In thermodynamics the concept of equilibrium is extended to include possible changes in the internal state of a system, as characterized by its temperature, pressure, density, and any other quantities needed to specify its state completely. At strict thermodynamic equilibrium, the temperature of the system is uniform (otherwise heat would flow), and any gradients in state functions such as pressure or density are balanced by external forces so that they remain constant. For example, the equilibrium pressure at the bottom of a column of air is higher than at the top because of the force of gravity, and density gradients in a centrifuge are balanced by the centrifugal force. It is also useful to consider quasi-equilibrium processes where, for example, temperature gradients are allowed if the rate of heat flow is too slow to be significant (adiabatic processes), but the system is otherwise in local thermodynamic equilibrium. For example, the adiabatic expansion of a rising column of air accounts for the decrease of atmospheric temperature with altitude.

Thermal Equilibrium

Heat is the flow of energy from a high temperature to a low temperature. When these temperatures balance out, heat stops flowing, then the system (or set of systems) is said to be in thermal equilibrium. Thermal equilibrium also implies that there›s no matter flowing into or out of the system. The zeroth law of thermodynamics uses thermal equilibrium to define how two different systems can be said to be at the same temperature. For example, when molten rock comes up from a volcano, it will give off heat to the atmosphere until the rock and the atmosphere are at the same temperature. Even though the two systems (rock and air) are very different, thermal equilibrium allows a definition of temperature for both.

All systems tend towards thermal equilibrium over time—and some systems will take a lot longer than others. Knowing that interacting systems will tend towards the same temperature allows for important applications in all areas of science. For example, the specific heat capacity of a substance can be determined by placing it in water and measuring the temperature after a period of time.

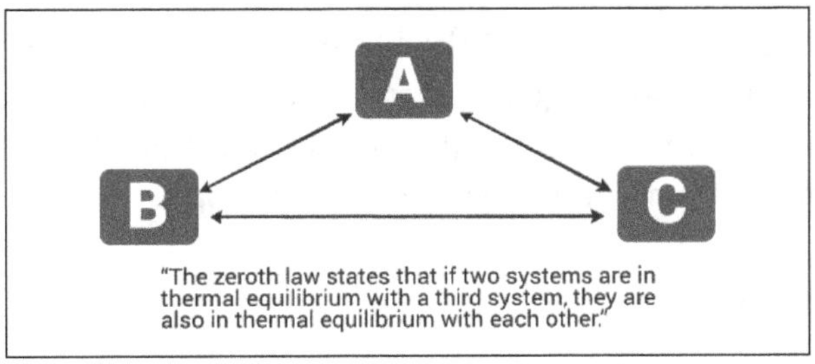

The zeroth law of thermodynamics.

Mechanical Equilibrium

A system is said to be in mechanical equilibrium when there is no work transfer and pressure difference, between the system and the surrounding.

Let's fix U_A, N_A and U_B, N_B, but allow V to vary (the membrane is insulating, impermeable for gas molecules, but its position is not fixed). Following the same logic, spontaneous "exchange of volume" between sub-systems will drive the system towards mechanical equilibrium (the membrane at rest). The equilibrium macropartition should have the largest (by far) multiplicity Ω (U, V) and entropy S (U, V).

In mechanical equilibrium:

$$\frac{\partial S_{AB}}{\partial V_A} = \frac{\partial S_A}{\partial V_A} + \frac{\partial S_B}{\partial V_A} = \frac{\partial S_A}{\partial V_A} - \frac{\partial S_B}{\partial V_B} = 0 \Rightarrow \frac{\partial S_A}{\partial V_A} = \frac{\partial S_B}{\partial V_B}$$

The volume-per-molecule should be the same for both sub-systems, or, if T is the same, P must be the same on both sides of the membrane.

for ideal gas $\left(\frac{\partial S}{\partial V}\right)_{U,N} = \frac{kN}{V} = \frac{P}{T}$

Relation between entropy and pressure:

$$P \equiv T\left(\frac{\partial S}{\partial V}\right)_{U,N} = \left(\frac{\partial S}{\partial V}\right)_{U,N} / \left(\frac{\partial S}{\partial U}\right)_{V,N}$$

$$P = T\left(\frac{\partial S}{\partial V}\right)_{U,N}$$

This definition implies that you are holding both the internal energy and the number of particles constant in taking the derivative. This can be applied to the expression for the entropy of a monoatomic ideal gas:

$$S = Nk\left[\text{In}\left(\frac{V}{N}\left(\frac{4\pi mU}{3Nh^2}\right)^{3/2}\right) + \frac{5}{2}\right]$$

$$P = T\frac{\partial}{\partial V}(Nk\ln V) = \frac{NkT}{V}$$

This relationship is just the ideal gas law. But the ideal gas law can be obtained from just Newton's laws, which give an expression for average pressure of a gas. That along with the kinetic temperature gives the form of the gas law. So, looking at it from the other direction, the ideal gas law offers confirmation of the relationship between pressure and entropy.

Thermodynamic Identity

Let's assume N is fixed: $dS = \left(\frac{\partial S}{\partial U}\right)_{N,V} dU + \left(\frac{\partial S}{\partial V}\right)_{N,U} dV$

In thermal equilibrium: $\left(\frac{\partial S}{\partial U}\right)_{V,N} = \frac{1}{T}$

In mechanical equilibrium: $\left(\frac{\partial S}{\partial V}\right)_{U,N} = \frac{P}{T}$

$$\Rightarrow dS = \frac{1}{T}dU + \frac{P}{T}dV$$

$$dU = TdS - PdV$$

This useful summary relationship called the thermodynamic identity makes use of the power of calculus and particularly partial derivatives. It may be applied to examine processes in which one or more state variables is held constant, e.g., constant volume, constant pressure, etc. The thermodynamic identity holds true for any infinitesmal change in a system so long at the pressure and temperature are well defined. It is presumed that the number of particles is constant (i.e., you are dealing with the same system before and after the change).

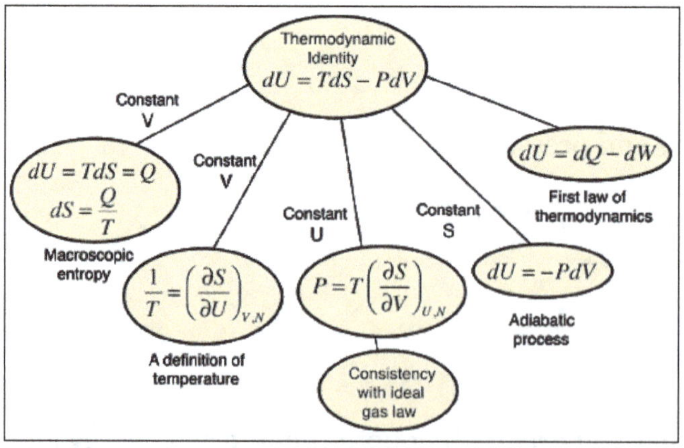

Quasistatic Processes

$$dU = TdS - PdV \quad \Leftrightarrow \quad dU = \delta Q + \delta W$$

(quasistatic processes with fixed N) (all processes)

Thus, for quasistatic processes:

$$\delta Q = TdS \qquad dS = \frac{\delta Q}{T}$$

We see that the above relation is valid even if the work is being done, provided that the pressure is well-defined and uniform throughout the system. In particular, for the adiabatic process,

$$\delta Q = 0 \Rightarrow \text{adiabatic} + \text{quasistatic} = \text{isentropic}$$

Comment on State Functions: $dS = \frac{\delta Q}{T}$ is an exact differential (S is a state function). Thus, the factor 1/T converts δQ into an exact differential for quasi-static processes.

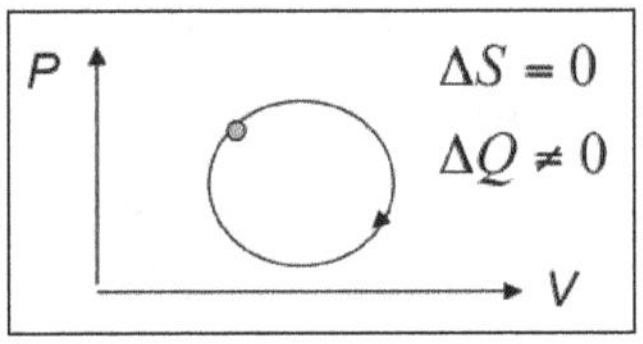

From the Sackur-Tetrode equation for an isentropic process:

$$\Delta S = 0 \Rightarrow VT^{f/2} = \text{const} \rightarrow VT^{\frac{1}{\gamma-1}} = \text{const}$$

Chemical Equilibrium

Chemical equilibrium is the thermodynamic equilibrium in a system where direct and reverse chemical reactions are possible. If chemical equilibrium takes place in the system, the rates of all reactions proceeding in two opposite directions are equal. Therefore, the macroscopic parameters of the system do not change and the relationship between concentrations of reacting substances remains constant at a given temperature. Equilibrium for any chemical reaction is expressed by an equality $\Sigma \nu_i \mu_i = 0$, where μ_i is the chemical potential of each reagent (i = 1,2,...) and ν_i is the stoichiometric coefficient of each substance in an equation of chemical reaction (it is positive for initial substances and negative for products of a reaction).

The dependence of chemical equilibrium on external conditions is expressed by the *Le Chatelier-Braun principle.* It consists of the following correlation: Let equilibrium take place and then influence the system, changing some external conditions (temperature, pressure, concentrations of reacting substances). The equilibrium of a reaction tends to follow such direction that allows the reduction of an external influence. A temperature increase will cause a displacement of the equilibrium to the direction of such reaction that proceeds with heat absorption. A pressure increase will cause equilibrium displacement to follow the direction of such reaction that leads to a volume decrease. The introduction of any additional reagent in the system will propel equilibrium displacement to a direction where this reagent is consumed.

The total *Gibbs energy* change of a chemical reaction aA + bB = cC + dD (when temperature and pressure are constant) is expressed by the equation:

$$\Delta_r G_{p,r} = \Delta_r G^o_{p,t} + RT \ \text{In}\left(\alpha_C^c \alpha_D^d \alpha_A^a \alpha_B^b\right),$$

where R is the gas constant, p is the pressure, T is the absolute temperature, α_i refers to the activities of the reacting substances and $\Delta_r G^o_{p,T}$ is the standard Gibbs energy's change of that reaction $(\alpha_i = 1)$. The value of $\Delta_r G^o_{p,T}$ can be calculated on the basis of standard values of the Gibbs energies of formation $(\Delta_f G^o)$ of the reagents at 298.15 K and of known thermodynamic relationships that determine the temperature and pressure dependencies of Gibbs energy change.

If equilibrium is attained, then,

$$\Delta_r G^o_{p,T} = -RT \, In \frac{(\alpha_{C equil})^c (\alpha_{D equil})^d}{(\alpha_{A equil})^a (\alpha_{B equil})^b} = -RT \, In \, K_\alpha .$$

Here, α_{equil} are the activities corresponding to the equilibrium state and K_a is the equilibrium constant expressed in terms of activities. Hence, it follows that $\Delta_r G_{P,T} = RT\left[In\left(\alpha_C^c \alpha_D^d / \alpha_A^a \alpha_B^b\right) - In \, K_\alpha\right]$. The last relationship is the van't Hoff isotherm equation (or *van't Hoff equation*). It permits the determination of a probable direction of the reaction under given conditions. The process will take place when $\Delta_r G_{P,T} < 0$, i.e., when $K_\alpha > \left(\alpha_C^c \alpha_D^d / \alpha_A^a \alpha_B^b\right)$. Analogous relationships can be obtained when the equilibrium constant (K_p) is expressed in terms of partial pressures (P_i) of the reagents:

$$\Delta_r G_{P,T} = RT\left[In \frac{P_C^c P_D^d}{P_A^a P_B^b} - In \, K_p\right]; \Delta_r G^o_{p,T} = -RT \, In \, K_p .$$

The "equilibrium constant of reaction" is the result of the mass action law, which determines a correlation between the masses of reacting substances under equilibrium. According to this law, the reaction's rate depends on the concentrations of reacting substances. The rate constant of a given reaction at fixed temperature is a constant value; therefore, the relationship of the rate constants of direct and reverse reactions is a constant value too. This relationship is a function of temperature only.

The equations that express a relationship between the $\Delta_r G^o_{p,T}$ -value and equilibrium constant of reaction allow the calculation of the equilibrium of chemical reactions, avoiding expensive and prolonged experiments. For such calculations, it is necessary to have reliable values of thermodynamic functions for all reacting substances.

Various experimental methods are used to determine equilibrium constants of chemical reactions. There are static and dynamic methods as well as the circulation method. The last is a specific combination of the static and dynamic methods. When static methods are used, the reaction mixture stays at a given temperature until an adjustment of the equilibrium takes place. Then "tempering" and chemical analysis of the reaction mixture are carried out. "Equilibrium tempering" is the fast-cooling of the reaction mixture to a low temperature where the rate of reaction is very small.

The more common dynamic method of defining equilibrium constants has often been called the transportation method. A steady stream of inert gas is passed over the mixture of substances that is maintained at a constant temperature. This "carrier" gas removes the volatile components of the reaction at a rate that depends on the rate of gas flow. The vapors of the reagents are condensed or collected by absorption or chemical combination at the colder portion of the apparatus. The experiments are carried out at different rates of gas flow. The equilibrium pressures of volatile reagents are determined by extrapolation of the results up to zero rate of the carrier gas.

A modification of the dynamic method used for investigating heterogeneous equilibria is the circulation method. The gas mixture is circulated in a closed space; circulation is carried out by means of electromagnetic pump. Equilibrium is attained when passing this mixture many times over the solid phase into the furnace. Tempering of the gas mixture is done when it is taken out from the hot zone and passed through a capillary. In view of the large linear rate of gas flow, this mixture becomes cold rapidly and its composition is not changed.

The most direct way of measuring equilibrium constants of chemical reactions is through the measurement of electromotive forces (the e.m.f. method). For example, the reaction:

$$Zn(cryst.) + CuSO_4(solution) = ZnSO_4(solution) + Cu(cryst)$$

is a process of potential generation for the Daniel galvanic element: $Zn^o/Zn^{2+}//Cu^o/Cu^{2+}$ A zinc plate (one electrode) is immersed into a solution of zinc sulfate and a copper plate (the other electrode) is immersed into a solution of copper sulfate. A galvanic element (source of electromotive force) can be created if both electrodes are connected by a tube that contains a solution-conductor. The dissolution of zinc (process: $Zn^o = Zn^{2+} + 2e$) takes place at one electrode; the precipitation of copper (process: $Cu^{2+} + 2e = Cu^o$) takes place at the second electrode. Therefore, the common potential forming reaction is: $Zn^o + Cu^{2+} = Zn^{2+} + Cu^o$. The Gibbs energy change for such reaction is given by the formula $\Delta_r G^o_T = -nFE_T$, where n is the number of gramme-equivalents of reagent; F is Faraday's constant (nF is the number of coulombs of electricity passed); and E_T is the electromotive force of the galvanic element at a given temperature. The value of the Gibbs energy of reaction can be used for calculating its equilibrium constant (K):

$$\text{In } K = -\Delta_r G^o_T / RT = nFE_T / RT.$$

The equilibrium state is a thermodynamic state of a system that is permanent in time. This invariability is not connected with some external process taking place. There are different kinds of equilibria. If the equilibrium is "steady," then any adjacent states of the system are less steady. It would be necessary to spend external work for transition from the equilibrium state to these adjacent states. It is also typical that steady equilibrium can be approached from two opposite directions. It is attained when the rates of direct and reverse reactions are equal, but not under conditions when the process is stopped in general. The equality $dG = 0$ is a general condition for "steady" and "unsteady" equilibria, but the value of the second differential of Gibbs energy is positive under steady equilibrium ($d^2G > 0$) and negative under unsteady equilibrium ($d^2G < 0$). The conditions of stability of the equilibrium can be deduced using the second law of thermodynamics. These are: 1) The pressure increases at a constant temperature if volume decreases [$(dP/dV)_T < 0$]; and 2) the value of heat capacity is positive ($C_p > 0$).

The degree of stability of the different states of chemical systems can vary. States which possess some relative stability are called "metastable" states. Such states have often arisen due to kinetic factors, which create difficulties for the transition of a system from the metastable (unsteady) state to a steady equilibrium state.

Basic Postulates of Equilibrium Thermodynamics: Phenomlogical Way of looking at Macroscopic Systems

The fundamental problem which needs to be formulated for a given macroscopic system in equilibrium thermodynamics is how to determine the final equilibrium state of a macroscopic system and describe it. This problem is addressed through a set of basic postulates of Thermodynamics, which cannot be proved from some known laws but must be taken as fundamental axioms whose validity lies in their applications predictions matching with experimental results pointing towards its phenomenological methodology. Incidentally, thermodynamic or macroscopic state of a system can be fully described in terms of a very small number of thermodynamic variables with introduction of additional number of thermodynamical variables to take care of complexities as and when these arise. These postulates are:

- First Postulate: The equilibrium states of a macroscopic system can be fully described in terms of a set of extensive variables including its internal energy U. The other extensive parameters being volume V and number of particles in the system N or more, as per the typical thermodynamic system.
- Second Postulate: There exists an extensive parameter/variable called entropy, which is a function of theses extensive parameters i.e. $S = S(U, V, N)$ which is defined for all equilibrium states of the system. This function contains all the information of the system and is a fundamental equation of the system. It has following properties:
    - S is continuous and differentiable and is a monotonically increasing function of internal energy, provided all other extensive variables are kept constant. This implies that $\frac{\partial S}{\partial U} > 0$ (which actually means that temperature T > 0).
    - The entropy S of a composite system (a system made up of macroscopic subsystems is an additive quantity implying thereby that it is an extensive parameter itself. i.e. $S = S_1 + S_2 + S_3 \dots\dots\dots\dots + S_N$. It also means that entropy is a homogeneous function of first degree in all its variables i.e. $S(\lambda U, \lambda V, \lambda N) = \lambda S(U, V, N)$.
- Third Postulate: If a macroscopic state is not in a state of equilibrium, its entropy will be less than the entropy of the equilibrium state with same values of extensive variables on which it depends. However, left to itself for the fixed values of extensive variables, a non-equilibrium state shall gain entropy to reach the equilibrium state implying that entropy will be maximum for the equilibrium state. This postulate also states that entropy of a system is zero in the state where, $\left(\frac{\partial U}{\partial S}\right)_{V,N} = 0$ i.e. T = 0. This is also called Nernst's Theorem.

Fundamental Thermodynamic Relations and Intensive Variables

To see this connection we must appreciate the first and second postulates together, which define a fundamental relation S = S (U, V, N) where we limit ourselves to only three extensive variables (the simplest possible system). But this relationship is not unique in the sense that we solve this to write U = U(S, V, N) or V = V (S, U, N) or N = N (S, V, U) each being a homogeneous function of first order. However, we shall confine to only the first two relations known as entropy representation and energy representation respectively and see how intensive variables can be defined from these relations.

First let us look at the energy representation and write down it in differential form i.e.

$$dU = \left(\frac{\partial U}{\partial S}\right)_{V,N} dS + \left(\frac{\partial U}{\partial V}\right)_{S,N} dV + \left(\frac{\partial U}{\partial N}\right)_{S,V} dN$$

Where, subscripts represent the variables which are to be kept constant while taking the partial derivatives.

In terms of conservation of energy this change in internal energy must be equal to heat, mechanical work performed for compression, and chemical energy resulting from raising the number of particles:

$$dU = TdS - PdV + \mu dN$$

Leading to following definitions of intensive variables:

- Temperature:

$$T = \left(\frac{\partial U}{\partial S}\right)_{V,N}$$

- Pressure:

$$P = -\left(\frac{\partial U}{\partial V}\right)_{S,N}$$

- Chemical potential:

$$\mu = \left(\frac{\partial U}{\partial N}\right)_{S,V}.$$

The functions $T = T(S, V, N)$, $P = P(S, V, N)$ and $\mu = \mu(S,V,N)$ are the equations of state in this representation. These equations express an intensive variable in terms of extensive variables. For example, from an ideal gas equation $pV = Nk_BT$, we get $P = \frac{Nk_BT}{V}$, law of equipartition $U = \frac{3}{2}NK_BT$ for an ideal gas leads to another equation of state $\frac{1}{T} = \frac{\frac{3}{2}(Nk_B)}{U}$

On the other hand in the entropy representation, we have:

$$dS = \left(\frac{\partial S}{\partial U}\right)_{V,N} dU + \left(\frac{\partial S}{\partial V}\right)_{U,N} dV + \left(\frac{\partial S}{\partial N}\right)_{U,V} dN$$

$$dS = \frac{1}{T}dU + \frac{P}{T}dV - \frac{\mu}{T}dN$$

Giving intensive variables as,

$$\frac{1}{T}=\left(\frac{\partial S}{\partial U}\right)_{V,N}$$

$$\frac{P}{T}-\left(\frac{\partial S}{\partial V}\right)_{U,N}$$

$$-\frac{\mu}{T}=\left(\frac{\partial S}{\partial N}\right)_{U,V}$$

$P=P(U,V,N), T=(U,V,N)$ and $\mu=\mu(U,V,N)$ are homogeneous functions of degree zero such that $\frac{P}{T}(\lambda U,\lambda V,\lambda N)=\frac{P}{T}(U,V,N), \frac{1}{T}(\lambda U,\lambda V,\lambda N)=\frac{1}{T}(U,V,N)$ and $\frac{\mu}{T}(\lambda U,\lambda V,\lambda N)=\frac{\mu}{T}(U,V,N)$.

The above result can be extending further by identifying other pairs of conjugate variables as stated earlier, whose product gives us energy e.g. velocity v and momentum p, angular velocity ω and angular momentum L, electrifield $\mathbb{E}$ and Polarisation P, magnetic field H and magnetization M leading to,

$$dU = TdS - PdV + \mu dN + vdp + \omega dL + \mathbb{E}dp + HdM$$

For multi component system, containing more than one type of particles the above equation becomes,

$$dU = TdS - PdV + \sum_i \mu_i dN_i + vdp + \omega dL + \mathbb{E}dP + HdM$$

Euler Relation and Gibbs Duhem Relation

From the fact that internal energy U(S, V, N) is an extensive quantity we can derive two useful relations.

Extensivity implies that $U(\lambda S,\lambda V,\lambda N)=\lambda U(S,V,N)$. Taking the derivative of both sides with respect to λ, we get,

$$\frac{\partial U(\lambda S,\lambda V,\lambda N)}{\partial(\lambda S)}S+\frac{\partial U(\lambda S,\lambda V,\lambda N)}{\partial(\lambda V)}V+\frac{\partial U(S,V,N)}{\partial(N)}N=U(S,V,N)$$

Put $\lambda=1$

$$\frac{\partial U(S,V,N)}{\partial(S)}S+\frac{\partial U(S,V,N)}{\partial(V)}V+\frac{\partial U(S,V,N)}{\partial(N)}N=U(S,V,N)$$

Which is Euler relation,

$$TS-PV+\mu N=U.$$

Differential form of this relation gives,

$$TdS - pdV + +\mu dN + SdT - VdP + N\,d\mu = dU.$$

This is true only if $SdT - VdP + N\,d\mu = 0$, i.e.

$$d\mu = \frac{V}{N}dP - \frac{S}{N}dT$$

This relation is known as Gibbs-Duhem relation.

Thermodynamic Potentials

Thermodynamic potentials are state functions that, together with the corresponding equations of state, describe the equilibrium behavior of a system as a function of so-called "natural variables". The natural variables are a set of appropriate variables that allow to compute other state functions by partial differentiation of the thermodynamic potentials.

Internal Energy U

The basic relation of thermodynamics is given by the equation:

$$dU = TdS + \sum_{i=1}^{m} F_i dq_i + \sum_{j=1}^{\alpha} \mu_j dN_j$$

where {F, q} denote the set of conjugate intensive and extensive variables that characterize a system. For instance, for a gas:

$$\{F,q\} \rightarrow \{-P,V\},$$

and for a magnetic system,

$$\{F,q\} \rightarrow \{\mathcal{B},\mathcal{M}\}.$$

Chemical Potential

The number of particles in the system is a natural extensive variable for the free energy, we did keep it hitherto constant. The number of particle of a distinct types j is denoted by N_j in ($dU = TdS + \sum_{i=1}^{m} F_i dq_i + \sum_{j=1}^{\alpha} \mu_j dN_j$), where j = 1,...,**α**. The respective intensive variable, μ (and respectively the μ_j), is denoted the chemical potential.

The chemical potential become identical to the Fermi energy for a gas of Fermions (at low temperatures,). Note that diffusion processes (e.g. across a membrane) may change individual N_j. The number of particles needs therefore not to be constant.

Internal Energy of a Gas

Equation $dU = TdS + \sum_{i=1}^{m} F_i dq_i + \sum_{j=1}^{\alpha} \mu_j dN_j$ is equivalent to,

$$dU = TdS - PdV + \mu dN, \quad U = U(S,V,N),$$

for a gas with one species of particle, which implies that,

$$T = \left(\frac{\partial U}{\partial S}\right)_{V,N}, \quad -P = \left(\frac{\partial U}{\partial V}\right)_{S,N}, \quad \mu = \left(\frac{\partial U}{\partial N}\right)_{S,V}.$$

Response Functions

The experimentally important response functions are obtained by second-order differentiation of the internal energy,

$$\left(\frac{\partial^2 U}{\partial S^2}\right)_{V,N} = \left(\frac{\partial T}{\partial S}\right)_{V,N} = \frac{T}{C_V}, \; C_V = T\left(\frac{\partial S}{\partial T}\right)_{V,N} = T\left[\left(\frac{\partial^2 U}{\partial S^2}\right)_{V,N}\right]^{-1}$$

$$\left(\frac{\partial^2 U}{\partial V^2}\right)_{S,N} = -\left(\frac{\partial P}{\partial V}\right)_{S,N} = \frac{1}{V\kappa_S}, \; \kappa_S = \frac{1}{V}\left[\left(\frac{\partial^2 U}{\partial V^2}\right)_{S,N}\right]^{-1}$$

Maxwell Relations

A Maxwell relation follows, from the differentiability of thermodynamic potentials. An example of a Maxwell relation derived from the differential of the internal energy U = U(S, V, N) is,

$$\frac{\partial}{\partial V}\left(\frac{\partial U}{\partial N}\right) = \frac{\partial}{\partial N}\left(\frac{\partial U}{\partial V}\right), \quad \left(\frac{\partial \mu}{\partial V}\right)_S = -\left(\frac{\partial P}{\partial N}\right)_V$$

which relates the change of the chemical potential μ with the volume V to the (negative of the) change of the pressure P with the number of particles N.

Monoatomic Ideal Gas

For the monoatomic ideal we did find hitherto,

$$PV = nRT, \quad U = \frac{3}{2}nRT, \quad C_V = \frac{3}{2}nR.$$

We now use these relations to derive an expression for U in terms of its natural variables.

Entropy. We start with the entropy:

$$S(T,V) - S(T_o,V_o) = \frac{3nR}{2}\log\left(\frac{T}{T_o}\right) + nR\log\left(\frac{V}{V_o}\right)$$

$$= \frac{3nR}{2}\log\left(\frac{U}{U_o}\right) + nR\log\left(\frac{V}{V_o}\right)$$

of the ideal gas, where we have used the ideal gas relation U = 3nRT/2 to substitute T and T_o with U and U_o in the second step.

With C_V = 3nR/2 we may write equivalently,

$$\frac{S-S_o}{C_V} = \text{In}\left(\frac{U}{U_o}\right) + \frac{2}{3}\text{In}\left(\frac{V}{V_o}\right) = \text{In}\left[\frac{U}{U_o}\left(\frac{V}{V_o}\right)^{2/3}\right],$$

which can be solved for the internal energy as,

$$U(S,V) = U_o\left(\frac{V_o}{V}\right)^{\gamma-1} e^{(S-S_o)/C_V}, \quad \gamma = 5/3.$$

Equation above is the fundamental equation for the ideal gas, with U(S, V) as the thermodynamic potential. S, V are the independent natural variables.

Thermodynamic Potential vs. Equation of State

The natural variables for U are S and V, which means that if the function U(S, V) is known for a given system we can obtain –all– thermodynamic properties of the system through the differentiation of U(S, V). The equation of state:

$$U = U(T, V, N)$$

for the internal energy U is on the contrary –not– a thermodynamic potential. This is because the first derivatives of above equation yield the specific heat C_V.

$$\left(\frac{\partial U}{\partial T}\right)_V = C_V, \quad \left(\frac{\partial U}{\partial V}\right)_T = T\left(\frac{\partial P}{\partial T}\right)_V - P,$$

but not the dependent variables S and P. Note the marked difference to,

$$(T = \left(\frac{\partial U}{\partial S}\right)_{V,N}, \quad -P = \left(\frac{\partial U}{\partial V}\right)_{S,N}, \quad \mu = \left(\frac{\partial U}{\partial N}\right)_{S,V}).$$

Classical Mechanics vs. Thermodynamics

There is a certain analogy between the potential V and internal energy U of classical mechanics and thermodynamics respectively:

	Classical mechanics	Thermodynamics
Potential	$V(x,y,z)$	$U(S, V, N)$

Independent variables	(x,y,z)	S, V, N
Dependent variables	$F_x = -\frac{\partial V}{\partial x}$	$P = -\left(\frac{\partial U}{\partial V}\right)_S$

Legendre Transformation

A disadvantage of using U(S, V, N) as a thermodynamic potential is that the natural variable S is difficult to control in the lab. For practical purposes, it is more convenient to deal with other thermodynamic potentials that can be defined by making use of the Legendre transformation.

Legendre transformations in classical mechanics. We recall the Legendre transformation connecting the Legendre function L(q, q˙) with the Hamilton function H(q, p), where q is the generalized coordinated and, q˙ and p respectively the velocity and the momentum, with,

$$p = \left(\frac{\partial L}{\partial \dot{q}}\right)_q ,$$

at constant coordinate q, and

$$H(q,p) = p\dot{q} - L(q,\dot{q}).$$

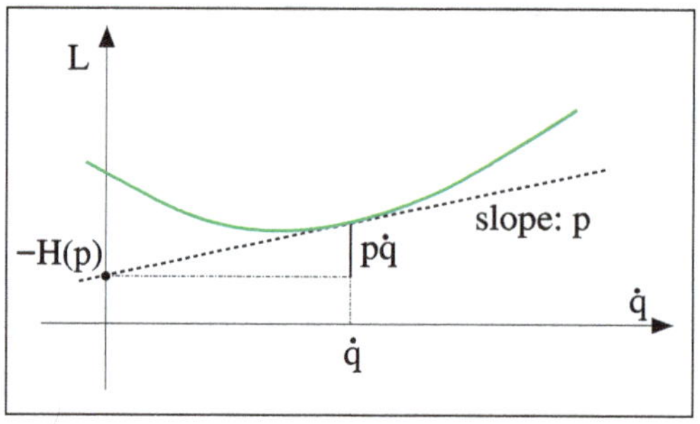

A Legendre transformation is hence a variable transformation, $\dot{q} \rightarrow p$, where one of the variables, here the momentum p, is defined by the slope of original function, viz by the slope of the Lagrange function $L(p,\dot{q})$.

Legendre Transformations in Thermodynamics

As an example we consider the transformation,

$$U(S,\ V,\ N) \rightarrow\ H(S,\ P,\ N),$$

from the internal energy U(S, V, N) to the enthalpy H(S, P, V). The form is,

$$H(S,\ P,\ N) = U(S,\ V,\ N) + PV,\ P = -\left(\frac{\partial U}{\partial V}\right)_S$$

The pressure is defined as the (negative) slope of the internal energy U with respect to the volume V, in analogy to the transformation from the velocity $\dot{q}$ to the momentum $p = \partial L / \partial \dot{q}$ in classical mechanics. Note however the different sign conventions between $(H(S, P, N) = U(S, V, N) + PV,\ P = -\left(\frac{\partial U}{\partial V}\right)_S)$ and $(H(q,p) = p\dot{q} - L(q,\dot{q}))$.

Differential

The Legendre transformation $(H(S, P, N) = U(S, V, N) + PV,\ P = -\left(\frac{\partial U}{\partial V}\right)_S)$ allows to evaluate with,

$$dH = d(U + PV) = (TdS - PdV + \mu dN) + (PdV + VdP),$$

the differential dH of the enthalpy in terms of its natural variables S, P and N. The result is,

$$dH = TdS + VdP + \mu dN.$$

The Four Basic Thermodynamic Potentials

A thermodynamic potential for a system with variable number of particles should depend on {μ, N} as well as a thermal variable and a mechanical variable, which can be (for a gas):

Thermal:	S (extensive)	T(intensive)
Mechanical:	V (extensive)	P (intensive)

The four possible combinations of these variables,

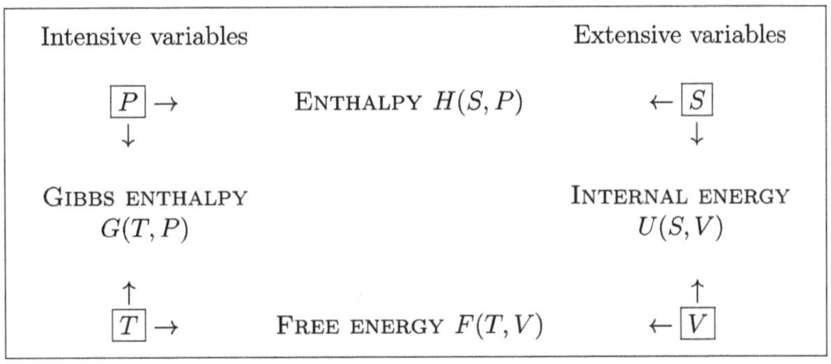

Four basic thermodynamic potentials:

- U(S, V, N) – Internal energy
- H(S, P, N) – Enthalpy
- G(T, P, N) – Gibbs enthalpy
- F(T, V, N) – Free energy

Gibbs Enthalpy

The differentials $(dU = TdS - PdV + \mu dN,\ \ U = U(S,V,N))$, dU = T dS–P dV +μdN, and

($dH = TdS + VdP + \mu dN$) of respectively the free energy U(S, V, N) and of the enthalpy H(S, P, N). For the Gibbs enthalpy one has in analogy,

$$G = U(S, V, N) - TS + PV, \qquad dG = -SdT + VdP + \mu dN.$$

Free Energy

The natural variables T, V and N of the free energy F = F(T, V, N) are straightforward to access experimentally. The free energy is hence an often considered thermodynamic potential. It obeys,

$$F = U(S, V, N) - TS, \qquad dF = -SdT - PdV + \mu dN.$$

Differential Relation between Free and Internal Energy

The Legendre transformation ($F = U(S, V, N) - TS, \quad dF = -SdT - PdV + \mu dN$) between U and F allow to derive a differential relation between the free and internal energy,

$$U = F + TS = F - T\left(\frac{\partial F}{\partial T}\right)_{V,N} = -T^2 \frac{\partial}{\partial T}\left(\frac{F}{T}\right),$$

which can be written in terms of the inverse temperature $\beta = 1/(k_B T)$ as,

$$U = \frac{\partial}{\partial \beta}(\beta F), \quad \frac{\partial}{\partial T} = \frac{\partial}{\partial(1/T)}\frac{\partial(1/T)}{\partial T} = \frac{-1}{T^2}\frac{\partial}{\partial(1/T)}.$$

It is generally easier to obtain explicit expressions for the internal energy U than for the free energy F. Above relation will be for this reason fundamental to the formulation of statistical mechanics, as it allows to identify F.

Maxwell Relations

The differentiability of the thermodynamic potentials leads to various Maxwell relations, such as the one given by Equation $\frac{\partial}{\partial V}\left(\frac{\partial U}{\partial N}\right) = \frac{\partial}{\partial N}\left(\frac{\partial U}{\partial V}\right), \quad \left(\frac{\partial \mu}{\partial V}\right)_S = -\left(\frac{\partial P}{\partial N}\right)_V$. There are furthermore four Maxwell relations relating T, respectively S to P respectively V. We drop in the following the dependence on the number of particles N.

U(S, V)	dU = TdS – PdV	$\left(\frac{\partial T}{\partial V}\right)_S = -\left(\frac{\partial P}{\partial S}\right)_V$
H(S, P)	dH = TdS + VdP	$\left(\frac{\partial T}{\partial P}\right)_S = \left(\frac{\partial V}{\partial S}\right)_P$
G(T, P)	dG = –SdT + VdP	$\left(\frac{\partial S}{\partial P}\right)_T = -\left(\frac{\partial V}{\partial T}\right)_P$
F(T, V)	dF = –SdT – PdV	$\left(\frac{\partial S}{\partial V}\right)_T = \left(\frac{\partial P}{\partial T}\right)_V$

Entropy

The differential of the entropy S = S(U, V, N) is obtained rewriting

$(dU = TdS - PdV + \mu dN, \quad U = U(S,V,N)$, as

$$dS = \frac{1}{T}dU + \frac{P}{T}dV - \frac{\mu}{T}dN.$$

The conjugate (natural) intensive variables are here 1/T and P/T with respect to dU and dV respectively. It is hence not customary to perform Legendre transformations with respect to these variables.

Homogeneity Relations

The thermodynamic potentials vary as extensive state properties if the volume and number of particles vary by a factor λ, i.e.

$$\left.\begin{array}{l} V \to \lambda V \\ N_j \to \lambda N_j \end{array}\right\} \Rightarrow U \to \lambda U.$$

Note that long-range interactions between particles, such as the unscreened Coulomb interaction, may invalidate above equation. Surface effects will contribute in addition for the case of a mesoscopic system.

Extensive and Intensive Variables

Temperature is intensive, but extensive:

$$dU = TdS - PdV + \sum_{j=1}^{\alpha} \mu_j dN_j.$$

It follows likewise that,

Extensive : U(S, V, N), H(S, P, N), G(T, P, N), F(T, V, N)

: S(U, V, N), V, N.

Intensive : T, P, μ.

Scaling

Doubling the volume V will also double the free energy F(T, V, N). All extensive functions, viz all thermodynamic potentials obey hence the scaling relations of the sort of,

$$\begin{aligned} F(T, \lambda V, \lambda N) &= \lambda F(T, V, N) \\ G(T, P, \lambda N) &= \lambda G(T, P, N) \end{aligned}$$

These functional dependencies express the homogeneity of the equilibrium thermodynamic state.

Gibbs-Duhem Relation

Scaling relations are generically of fundamental importance in physics. Here we use the extensivity of G and differentiate $\begin{matrix} F(T, \lambda V, \lambda N) & = \lambda F(T, V, N) \\ G(T, P, \lambda N) & = \lambda G(T, P, N) \end{matrix}$ with respect to λ (at λ = 1),

$$\lambda G(T,P,N) = \frac{d}{d\lambda} G(T,P,\lambda N_1,\lambda N_2,...)\Big|_{\lambda=1} = \sum_{j=1}^{\alpha} \underbrace{\left(\frac{\partial G}{\partial(\lambda N_j)}\right)_{T,P,N_{i\neq j}}\Bigg|_{\lambda=1}}_{=\mu_j} N_j$$

We obtain then with,

$$G(T,P,N) = \sum_{j=1}^{\alpha} \mu_j N_j \Rightarrow \text{Gibbs-Duhem relation.}$$

the Gibbs-Duhem relation.

Chemical Potential

When there is only one class of particles (α = 1),

$$G(T, P, N) = \mu N.$$

The chemical potential may hence be interpreted as Gibbs enthalpy per particle.

Representation of the Internal Energy

The Gibbs-Duhem relation ($G(T, P, N) = \mu N$) allows to rewrite the other thermodynamic potential. We recall, e.g. the Legendre transform ($G = U(S, V, N) - TS + PV, \quad dG = -SdT + VdP + \mu dN$),

$$G = U - TS + PV, \quad U = TS - PV + \mu N.$$

Note, however, that this representation is only seemingly simple.

Grand Canonical Potential

In a system that interchanges particles with a reservoir, the chemical potential is constant and the number of particles changes accordingly. In order to describe such a system, we introduce the grand canonical potential Ω(T, V, μ) as the Legendre transform of F(T, V, N), where the natural variable N is replaced by μ,

$$\Omega(T, V, \mu) = F(T, V, N) - \mu N, \quad \mu(T,V,N) = \left(\frac{\partial F}{\partial N}\right)_{T,V}.$$

The total differential of Ω is,

$$d\Omega = dF - \mu dN - Nd\mu, \quad d\Omega = -SdT - PdV - Nd\mu.$$

Through differentiation of Ω we can calculate S, P and N as,

$$S=-\left(\frac{\partial\Omega}{\partial T}\right)_{V,\mu},\ P=-\left(\frac{\partial\Omega}{\partial V}\right)_{T,\mu},\ N=-\left(\frac{\partial\Omega}{\partial\mu}\right)_{T,V}.$$

Thermal Equation of State

From the Gibbs Duhem relation ($G(T,\ P,\ N)=\mu N$), G = μN, it follows that,

$$\Omega=F-\mu N=G-PV-\mu N=-PV,\ F=G-PV$$

and hence that the thermal equation of state P = P(T, V) takes the from,

$$P=-\frac{\Omega(T,V,\mu)}{V}\Rightarrow \text{thermal equation of state.}$$

Differential Relation between Ω and U

With, ($U=F+TS=F-T\left(\frac{\partial F}{\partial T}\right)_{V,N}=-T^2\frac{\partial}{\partial T}\left(\frac{F}{T}\right)$), F = U – T S, we may write

($\Omega(T,\ V,\ \mu)=F(T,\ V,\ N)-\mu N,\quad \mu(T,V,N)=\left(\frac{\partial F}{\partial N}\right)_{T,V}$) as

$$\Omega=F-\mu N=U-TS-\mu N,\quad S=-\left(\frac{\partial\Omega}{\partial T}\right)_{V,\mu}.$$

Reanranging the terms then leads to,

$$U-\mu N=\Omega+TS=\Omega+TS=\Omega-T\left(\frac{\partial\Omega}{\partial T}\right)_{V,\mu},\ U=\mu N=\frac{\partial}{\partial\beta}(\beta\Omega),$$

where we have used in analogy to ($U=\frac{\partial}{\partial\beta}(\beta F),\ \frac{\partial}{\partial T}=\frac{\partial}{\partial(1/T)}\frac{\partial(1/T)}{\partial T}=\frac{-1}{T^2}\frac{\partial}{\partial(1/T)}$) that

$$1-T\frac{\partial}{\partial T}=-T^2\frac{\partial}{\partial T}\frac{1}{T}=\beta\frac{\partial}{\partial T}\beta,\quad \frac{\partial}{\partial T}=\frac{\partial}{\partial(1/T)}\frac{\partial(1/T)}{\partial T}=\frac{-1}{T^2}\frac{\partial}{\partial\beta}.$$

Equilibrium Conditions

Thermodynamic potentials dealt hitherto mostly with the equilibrium state. The second law of thermodynamics,

$$TdS\ \geq\ \delta Q=dU+PdV-\mu dN,$$

allows however also to study the impact of irreversible processes, that is the evolution of the system toward the equilibrium. We have used for above equation the differential form together with the generalized first law,

$$dU = \delta Q + \delta W + \mu dN, \qquad \delta W = -PdV,$$

where we have taken the δW appropriate for a gas.

Equilibrium Internal Energy

The entropy is an exact differential $dS = \delta Q/T$ at equilibrium. One can then consider situations like,

$$dU = TdS - PdV \Rightarrow \begin{cases} dU = -PdV & S = \text{const. (work)} \\ dU = TdS & V = \text{const. (heat)} \end{cases}$$

where selected state variables are constant.

Principle of Maximal Entropy

Isolated systems are defined by,

$$\left.\begin{aligned} \delta Q &= 0 \\ dV &= 0 \\ dN &= 0 \end{aligned}\right\} \Rightarrow dU = 0.$$

The second law ($TdS \geq \delta Q = dU + PdV - \mu dN$) tells us then that,

$$\Delta S \geq 0$$
$$dS = 0 \quad \text{in equilibrium .}$$

> In all irreversible processes, under the conditions:
>
> U = const., V = const., N = const.
>
> The entropy increases and is maximal in the stationary equilibrium.
>
> S evolves to a maximum under conditions of fixed U, V, N.

Two Isolated System

Consider two isolated systems 1 and 2 separated by a wall, which is however movable and which allows for the exchange of energy and particles. The variables V_1, V_2, U_1, and U_2, N_1, N_2 must obey the following constraints:

$$U = U_1 + U_2 = \text{const.} \Rightarrow dU_1 = -dU_2$$

$$V = V_1 + V_2 = \text{const.} \Rightarrow dV_1 = -dV_2$$

$$N = N_1 + N_2 = \text{const.} \Rightarrow dN_1 = -dN_2$$

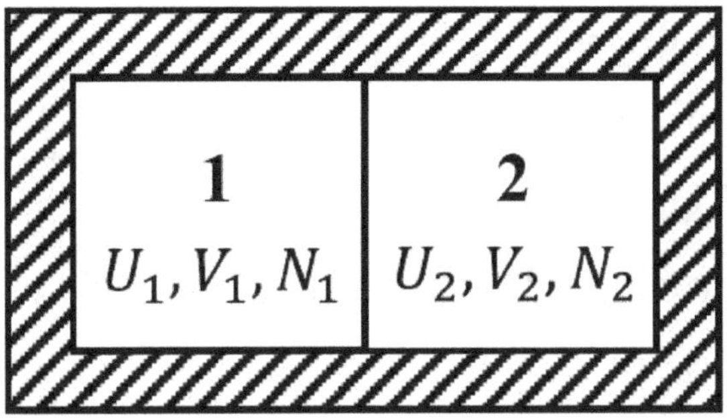

The total entropy of 1 and 2 is likewise a sum of the individual entropies:

$$S = S(U_1, V_1, N_1) + S(U_2, V_2, N_2) = S_1 + S_2.$$

Equilibrium conditions: Both systems will interchange particles and energy until equilibrium is attained. In the equilibrium state,

$$
\begin{aligned}
0 = dS &= dS_1 + dS_2 \\
&= \left\{ \left(\frac{\partial S_1}{\partial U_1}\right)_{V_1,N_1} - \left(\frac{\partial S_2}{\partial U_2}\right)_{V_2,N_2} \right\} dU_1 \\
&+ \left\{ \left(\frac{\partial S_1}{\partial V_1}\right)_{U_1,N_1} - \left(\frac{\partial S_2}{\partial V_2}\right)_{U_2,N_2} \right\} dV_1 \\
&+ \left\{ \left(\frac{\partial S_1}{\partial N_1}\right)_{U_1,V_1} - \left(\frac{\partial S_2}{\partial N_2}\right)_{U_2,V_2} \right\} dN_1 \\
&= \left(\frac{1}{T_1} - \frac{1}{T_2}\right) dU_1 + \left(\frac{P_1}{T_1} - \frac{P_2}{T_2}\right) dV_1 + \left(-\frac{\mu_1}{T_1} + \frac{\mu_2}{T_2}\right) dN_1.
\end{aligned}
$$

where we have used the steady-state version of the second law ($TdS \geq \delta Q = dU + PdV - \mu dN$), dS = dU + PdV − μdN.

Homogeneity Condition

U_1, V_1, and N_1 are independent variables. The prefactors of dU_1, dV_1, and dN_1 need hence to vanish individually:

$$T_1 = T_2 = T; \quad P_1 = P_2 = P; \quad \mu_1 = \mu_2 = \mu.$$

Instead of two systems we can subdivide a given system in an arbitrary number of ways. The system needs consequently to be homogeneous: An isolated system in equilibrium has everywhere the same temperature, pressure and chemical potential.

Principle of Minimal Free Energy

Closed systems in a reservoir with heat, but without work interchange, are defined by,

Closed $\Rightarrow N_j = \text{const} \Rightarrow dN_j = 0$,

In a reservoir $\Rightarrow T = \text{const} \Rightarrow dT = 0$,

Without work interchange $\Rightarrow V = \text{const} \Rightarrow dV = 0$.

The second law of thermodynamics ($TdS \geq \delta Q = dU + PdV - \mu dN$) reduces then to,

$$TdS \geq dU + PdV - \mu dN = dU, \qquad dV = 0 = dN.$$

Principle of minimal free energy. For constant temperature one has T dS = d(T S) and consequently that,

$$0 \leq TdS - dU = d(TS - U), \qquad dF = d(U - TS) \leq 0,$$

where we have used ($F = U(S,\ V,\ N) - TS, \quad dF = -SdT - PdV + \mu dN$), namely that F = U − T S.

The free energy decreases and reaches a minimum at equilibrium,

$dF \leq 0$

$dF = 0$ in equilibrium

for all irreversible processes characterized by a constant temperature T, volume V and particle number N.

Composite System

Consider a composite system in a heat reservoir at temperature T, whose particles can pass through the wall. It obeys the conditions,

$$V = V_1 + V_2 = \text{const.} \Rightarrow dV_1 = -dV_2$$
$$N = N_1 + N_2 = \text{const.} \Rightarrow dN_1 = -dN_2$$
$$F = F(T, V_1, N_1) + F = (T, V_2, N_2) = F_1 + F_2$$

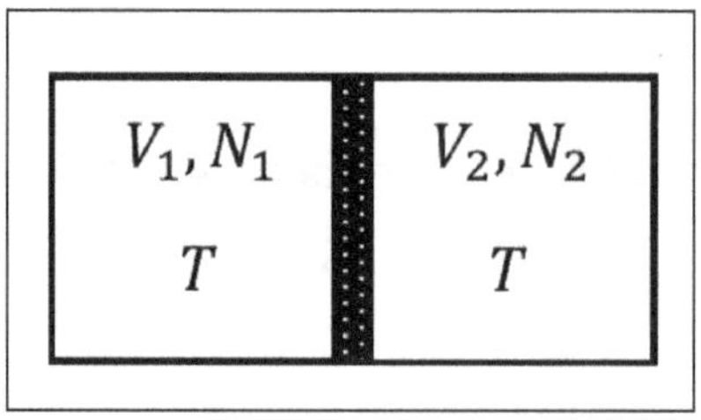

In equilibrium,

$$0 = dF = dF_1 + dF_2$$

$$= \left\{ \left(\frac{\partial F_1}{\partial V_1} \right)_{N_1,T} - \left(\frac{\partial F_2}{\partial V_2} \right)_{N_2,T} \right\} dV_1$$
$$+ \left\{ \left(\frac{\partial F_1}{\partial N_1} \right)_{V_1,T} - \left(\frac{\partial F_2}{\partial N_2} \right)_{V_2,T} \right\} dN_1$$
$$= \{-P_1 + P_2\} dV_1 + \{\mu_1 - \mu_2\} dN_1.$$

With V_1 and N_1 being independent variables if follows that,

$$P_1 = P_2 = P, \quad \mu_1 = \mu_2 = \mu.$$

A closed system with constant volume V which is in contact with a heat bath at constant temperature T reaches an equilibrium state characterized by a uniform pressure P and chemical potential μ.

Principle of Minimal Gibbs Energy

Generalizing above considerations we consider closed system in a heat reservoir, but at constant pressure P (previously V was constant). One can easily show that,

$$dT = 0, \ dN_j = 0, \ dP = 0$$

Implies that,

$$dG \le 0$$

$$dG = 0 \text{ in equilibrium.}$$

The Gibbs energy decreases in an irreversible process for a system with constant pressure P in contact with a heat bath at constant temperature T. G(T, P, N) reaches its minimum at equilibrium, which is characterized by a uniform chemical potential μ.

Stationarity Principles in Information Theory

From the microscopic point of view the equilibrium thermodynamic state is stationary in the sense that the positions and the velocities of the constituent particle change continuously. The ongoing reconfiguration of the microscopic state leaves however the macroscopic properties of the state intact.

Stationary probability distribution functions. A probability distribution functions p(y) quantifies the probability that the state y (an event) occurs, with,

$$\int p(y) dy = 1, \quad p(y) \ge 0.$$

It is the task of statistical mechanics to find the expression for p(y) Statistical mechanics dealing hence with probability distribution functions which are stationary with respect to the reconfiguration of microscopic states.

Entropy and Disorder

In statistical mechanics the entropy has, The form:

$$S[p]=-k_B\langle \text{In}(p)\rangle=-k_B\int p(y)\text{In}(p(y))dy.$$

It measures the amount of disorder in the sense that a flat (maximally disordered) distribution function p(y) has the largest entropy S[p]. Here we have written S = S[p] as a functional of p = p(y). Irreversible processes increase the amount of disorder in isolated systems. The entropy of the universe can only increase.

Entropy and Information

Information theory deals with information processing as described by probability distribution functions. The central quantity of information theory is the Shannon entropy,

$$I[p]=-\int p(y)\log(p(y))dy,$$

which is identical, apart from an overall factor, kB, to the thermodynamic entropy. The Shannon entropy is however used as a measure of the information content. The context is here information transmitted via communication channels.

Bit sequence	Type	Randomness	Information content
..01010101..	Ordered	Zero	Zero (predictable)
..00110101..	Disordered	High	High

Shannon's Source Coding Theorem

A precise connections between the informationtheoretical entropy I[p] and the information content of a communication channel is provided by the Shannon coding theorem, which states the I[p] provides a lower bound for lossless compression. Information is inevitably lost when compressing more.

Life as a Stationary State

Life is by definition a stationary state of matter which retains its functionality despite the continuously ongoing turnover of its constituent elements. The proteins, enzymes and cells of our body have lifetimes ranging typically from hours to months.

Equilibrium Principles in the Neurosciences

The stationary equilibrium of the brain with both the environment and the body is dependent on a continuous flux of information. Information theoretical principles are consequently of widespread use in the neurosciences. An example is the free energy principle, as proposed by Carl Friston. It suggests that the brain may be minimizing the information-theoretical analog of the thermodynamic free energy.

Statistical Foundation of Thermodynamics

Realizing the fact that macroscopic systems are made up of typically 6.023×10^{23} points to the enormity of the systems involved. Thermodynamic limit is a way to express this enormity and a natural way to express this by carrying out the limit number of particles $N \rightarrow \infty$ and volume $V \rightarrow \infty$, such that the particle density n remains fixed. Taking this limit amounts to saying that extensive properties of the system varies proportional to the size of the system i.e. N or V where as intensive properties remain independent of the size of the system. The process can be start with a finite system with a fixed density and keep on adding more and more such identical finite systems to it such that number density remains the same causing increase only in extensive parameters without change in the intensive parameters. Not going to thermodynamic limit in the mathematical sense can lead to serious problems with the strict compliance of the extensive nature of extensive parameters such as entropy.

Macroscopic State and Microscopic State

There are two ways to characterize a macroscopic system in equilibrium under consideration. These two ways are respectively macroscopic state and microscopic state of the system. A given macroscopic state corresponds to a large number of different microstates.

A macroscopic state also called a macrostate is described in terms of macroscopic variables such as total energy E, volume V and number of particles N. In describing a macrostate we are not at all bothered by the description of the properties of individual particles constituting the system such as their momentum, spin etc.

A microscopic state also called microstate is specified by describing the state of each particle in a given macrostate. Counting the number of microstates in a given macrostate is the most basic task to indulge in statistical mechanics.

To further appreciate the difference, let us take a system m ade up of N number of non-interacting particles with total energy E. This energy is the sum of the energies ε_i of individual particles, such that,

$$E = \sum_i n_i \varepsilon_i$$

Where n_i is the number of particles each with energy ε_i. Obviously,

$$N = \sum_i n_i$$

It may be recalled that ε_i are the single particle energies which have two distinct properties:

- These are quantized, leading to quantization of total energy E.
- Their value depends on the volume to which they are confined. However, it is worth noting that as volume increases the difference between the adjacent energy levels decrease rapidly and E goes from being a discreet variable to a continuous variable.

So macrostate of a system under consideration is defined by specification of N, V and E.

This macrostate can be realized at the microscopic level in many different possible ways known as microstates. In simple words, it implies that total energy E can be distributed among N particles constituting the system in large numer of different ways.

Quantum mechanically this amounts to specifying all possible independent solutions $\psi(r_1, r_2, \ldots\ldots\ldots r_N)$ of the N particle Schrodinger equation of the system under consideration with energy eigen value E.

One of the aims of statistical mechanical formulation is count the number of all the possible microstates. In this course this number shall be denoted by Ω, which is a function of macroscopic variables of the system, which in the present case are N, V and E. Therefore, this count is a number denoted by the function Ω(N, V,E). The value $\frac{1}{\Omega}$, gives the pro bability of ocuurance of any one of the states, which is as per the principle of equal apriori probability the same for each microstate. This $\frac{1}{\Omega}$, is also known as thermodynamic probability P_r.

The equilibrium state of the system then is the state for which function Ω(N, V, E) maximizes for each of its arguments. It is interesting to note that though a system corresponding to a given macrostate remains in a state of equilibrium, its microscopic states may keep on changing continuously.

Principle of Equal Apriori Probabilities

This principle is the back bone of the statistical mechanics for systems at equilibrium. It states that all allowed microstates compatible with a given macrostate, (i.e. subject to constraints imposed by equations $E = \sum_i n_i \varepsilon_i$ and $N = \sum_i n_i$ or any other constaint on macroscopic variables involved) are equally likely.

Allowing possibility of all microstates of a given macrostate has resulted in derivation of ideal gas laws and derivation of Maxwell Boltzmann velocity distribution law which were later confirmed experimentally. Justification of this principle, therefore, lies in the results derived from it and observed experimentally. This principle is treated as one of the postulates of statistical mechanics.

Linking Thermodynamics with Statistics

We now proceed to establish a link between thermodynamics with statistics based on the the state that the equilibrium state of the system then is the state for which function Ω(N,V, E) maximizes for each of its arguments leading to three conditions of equilibrium for each of the microscopic variable E, V and N.

For this we consider two physical systems: system 1 and system 2 which are separately in equilibrium, described by the macroscopic variables N_1, V_1, E_1 and N_2. V_2, E_2. System 1 has Ω_1 (N_1, V_1, E_1) number of microstates and system 2 has Ω(N_2, V_2, 2) as shown in figure.

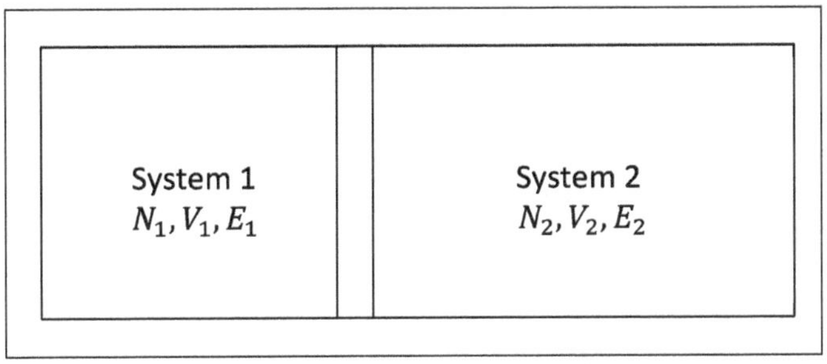

Two systems ready for making contact in three ways (a) when the partition is conductiong (b) when the partition is made conducting as well as movable and (c) when the partition also allows for exchange of particles.

Case: Thermal Contact Between Two Systems- Notion of Temperature in Statistical Mechanics: Let us consider first the case in which N_1, V_1 and N_2 V_2 are kept constant but energy can be exchanged by making the intervening partition conducting. However, in this case the total energy $E = E_1 + E_2$ remains conserved i.e.

$$E = E_1 + E_2 = \text{Constant}$$

At any instant system 1 is equally likely to be in any one of the microstates Ω_1 (N_1, V_1, E_1) and system 2 is equally likely to be in any one of the microstates Ω_2 (N_2, V_2, E_2). Then the combined system is equally likely to be any one of the microstates Ω(N, V, E), such that,

$$\begin{aligned}\Omega(N,V,E) &= \Omega_1(N_1,V_1,E_1)\Omega(N_2,V_2,E_2)\\ &= \Omega(N_1,V_1,E_1)\Omega(N_2,V_2,E-E_2)\end{aligned}$$

Since N_1, V_1 and N_2, V_2 are constant, we can write above equation as:

$$\Omega(E) = \Omega(E_1)\Omega(E-E_1)$$

Ω number of microstates of the combined system here vary with E_1 till equilibrium is established, and this happens until Ω maximizes. The combined system then attains a macrostate which has the most number of microstates. The important point to note is that this macrostate of the system is that state in which the system spends the largest portion of its time and presents itself as an equilibrium state.

Mathematical this condition when applied to ($\Omega(E) = \Omega(E_1)\Omega(E-E_1)$) means,

$$\frac{\partial\Omega(E,E_1)}{\partial E_1} = 0, \text{ when } E_1 = \overline{E_1}, \text{ the equilibrium value of } E_1 \text{ and } E_2 = \overline{E_2}.$$

We, therefore, obtain:

$$\left(\frac{\partial\Omega_1(E_1)}{\partial E_1}\right)_{E_1=\overline{E_1}}\Omega_2\left(\overline{E_2}\right)+\Omega_1\left(\overline{E_1}\right)\left(\frac{\partial\Omega_2(E_2)}{\partial(E_2)}\right)_{E_2=\overline{E_2}}\frac{\partial(E_2)}{\partial E_1} = 0$$

Note $E_2 = E - E_1$

$$\left(\frac{\partial\Omega_1(E_1)}{\partial E_1}\right)_{E=\overline{E_1}}\Omega_2\left(\overline{E_2}\right)-\left(\frac{\partial\Omega_2(E_2)}{\partial(E_2)}\right)_{E_2=\overline{E_2}}\Omega_1\left(\overline{E_1}\right)=0$$

$$\frac{1}{\Omega_1\left(\overline{E_1}\right)}\left(\frac{\partial\Omega_1(E_1)}{\partial E_1}\right)_{E=\overline{E_1}}=\frac{1}{\Omega_2\left(\overline{E_2}\right)}\left(\frac{\partial\Omega_2(E_2)}{\partial(E_2)}\right)E_2=\overline{E_2}$$

$$\left(\frac{\partial\ \mathrm{In}\ \Omega_2(E_1)}{\partial E_1}\right)_{E_1=\overline{E_1}}=\left(\frac{\partial\ \mathrm{In}\ \Omega_2(E_2)}{\partial E_2}\right)_{E_2=\overline{E_2}}$$

This equation is the condition for equilibrium of the system 1 and system 2.

Where we define $\beta=\left(\frac{\partial\ \mathrm{In}\Omega(N,E,V)}{\partial E}\right)E=\overline{E},V,N.$

So equation ($\left(\frac{\partial\Omega_1(E_1)}{\partial E_1}\right)_{E=\overline{E_1}}\Omega_2\left(\overline{E_2}\right)-\left(\frac{\partial\Omega_2(E_2)}{\partial(E_2)}\right)_{E_2=\overline{E_2}}\Omega_1\left(\overline{E_1}\right)=0$

$$\frac{1}{\Omega_1\left(\overline{E_1}\right)}\left(\frac{\partial\Omega_1(E_1)}{\partial E_1}\right)_{E=\overline{E_1}}=\frac{1}{\Omega_2\left(\overline{E_2}\right)}\left(\frac{\partial\Omega_2(E_2)}{\partial(E_2)}\right)E_2=\overline{E_2}$$

$$\left(\frac{\partial\ \mathrm{In}\ \Omega_2(E_1)}{\partial E_1}\right)_{E_1=\overline{E_1}}=\left(\frac{\partial\ \mathrm{In}\ \Omega_2(E_2)}{\partial E_2}\right)_{E_2=\overline{E_2}}\ )$$

implies,

$$\beta_1=\beta_2$$

What have we achieved? We found that:

- When two systems are brought in contact and allowed to exchange energy through conducting partition, the exchange of energy continues till equilibrium values $E=\overline{E}_1$ and $E_2=\overline{E}_2$ of the two systems is not attained. Once this happens systems come to thermal equilibrium.
- This stage of thermal equilibrium is indicated when the respective values of the parameter β,β_1 and β_2 become equal, pointing towards zeroth law of thermodynamics which defines temperature as that parameter. It is, therefore, natural to look for this connection between β and temperature T.
- If we recall that the temperature is defined by the derivative of the entropy with respect to energy E:

$$\left(\frac{\partial S}{\partial E}\right)_{N,V}=\frac{1}{T}$$

On comparing equations $\beta = \left(\frac{\partial \ln\Omega(N,E,V)}{\partial E}\right) E = \bar{E}, V, N$ and $\left(\frac{\partial S}{\partial E}\right)_{N,V} = \frac{1}{T}$ we find that

$$\frac{\Delta S}{\Delta(\ln\Omega)} = \frac{1}{\beta T} = \text{Constant}$$

Equation above was first established by Ludwig Boltzmann. He also underlined the connection between the two approaches i.e. thermodynamic approach, equation $\left(\frac{\partial S}{\partial E}\right)_{N,V} = \frac{1}{T}$ and statistical approach equation $\beta = \left(\frac{\partial \ln\Omega(N,E,V)}{\partial E}\right) E = \bar{E}, V, N$ and further concluded that the constant in equation $\frac{\Delta S}{\Delta(\ln\Omega)} = \frac{1}{\beta T} = \text{Constant}$ should be a universal constant, which was later named as Boltzmann constant by Planck as,

$$S = k_B \ln \Omega$$

Also,

$$\frac{1}{\beta T} = k_B \text{ or } \beta = \frac{1}{k_{BT}}$$

- According to formula ($S = k_B \ln \Omega$), one can get entropy of a system by knowing the total number of microstates of a macrostate for a given system.
- If Ω = 1, this corresponds to a unique completely ordered state, with entropy S = 0. As Ω increases, the number of microstates increase and more is the number of microstates more is the entropy. Therefore, in a given system disorder is manifested through the count of the microstates of a given macrostate of the system, larger this number larger is the disorder and less the degree of predictability.
- Furthermore, maximization of the number of possible microstates or probability amounts to maximizing the entropy for equilibrium.

Case: Mechanical Contact Between Two Systems- Notion of Pressure in Statistical Mechanics.

If we extend the considerations of the case and further relax the constraint on the partition separating system 1 from system 2 by making it movable as well as conducting figure, then the volumes of the two systems also become variables, though the total volume $V = V_1 + V_2$, remains constant and besides equation $\left(\frac{\partial \ln \Omega_2(E_1)}{\partial E_1}\right)_{E_1=\bar{E_1}} = \left(\frac{\partial \ln \Omega_2(E_2)}{\partial E_2}\right)_{E_2=\bar{E_2}}$ corresponding to thermal equilibrium,

$$\left(\frac{\partial \ln \Omega_2(E_1)}{\partial E_1}\right)_{N_1,V_1,E_1=\bar{E_1}} = \left(\frac{\partial \ln \Omega_2(E_2)}{\partial E_2}\right)_{N_2,V_2,E_2=\bar{E_2}}$$

We have another equation which can be obtained in an analogous way as in case giving another condition for the system to be in mechanical equilibrium as well:

$$\left(\frac{\partial \text{ In } \Omega_2(E_1)}{\partial V_1}\right)_{N_1,E_1,V_1=\overline{V_1}} = \left(\frac{\partial \text{In}\Omega_2(V_2)}{\partial V_2}\right)_{N_2,E_2,V_2=\overline{V_2}}.$$

And we can define another parameter η,

$$\eta = \left(\frac{\partial \text{ In } \Omega(N,E,V)}{\partial V}\right)_{V=\overline{V},E,N}$$

On comparing with, again recalling thermodynamic formula:

$$\left(\frac{\partial S}{\partial V}\right)_{N,E} = \frac{P}{T}$$

Equation $\left(\frac{\partial \text{ In } \Omega_2(E_1)}{\partial V_1}\right)_{N_1,E_1,V_1=\overline{V_1}} = \left(\frac{\partial \text{In}\Omega_2(V_2)}{\partial V_2}\right)_{N_2,E_2,V_2=\overline{V_2}}$ implies that,

$$\frac{\Delta S}{\Delta(\text{In } \Omega)} = \frac{P}{\eta T} = \text{Constant}$$

Therefore, we can conclude:

- The equilibrium condition during thermal and mechanical contacts, such that $\frac{P}{\eta T} = k_B$ or $\eta = \frac{P}{K_B T}$ becomes,

$$\frac{P_1}{T_1} = \frac{P_2}{T_2}$$

i.e. $T_1 = T_2$ and $P_1 = P_2$.

Case: Chermical Contact Between Two Systems- Notion of Chemical Potential in Statistical Mechanics

Next if we consider besides thermal contact, partition such that it not only allows thermal contact but also allows exchange of particles between system 1 and system 2, but keep the volume fixed. Then only total Number of particles $N = N_1 + N_2$ and total Energy $E = E_1 + E_2$ remain fixed but V_1, V_2, E_1 and E_2 of the two systems can change. So besides equation $\left(\frac{\partial \text{ In } \Omega_2(E_1)}{\partial E_1}\right)_{E_1=\overline{E_1}} = \left(\frac{\partial \text{ In } \Omega_2(E_2)}{\partial E_2}\right)_{E_2=\overline{E_2}}$.

$$\left(\frac{\partial \text{ In } \Omega_1(E_1)}{\partial E_1}\right)_{N_1,V_1,E_1=\overline{E_1}} = \left(\frac{\partial \text{ In } \Omega_2(E_2)}{\partial E_2}\right)_{N_2,V_2,E_2=\overline{E_2}}$$

We have another equation which can be obtained in an analogous way as in case giving another condition for the system to be in chemical equilibrium as well,

$$\left(\frac{\partial \text{ In } \Omega_1(N_1)}{\partial N_1}\right)_{N_1=\overline{N_1},E_1V_1} = \left(\frac{\partial \text{ In } \Omega_2(N_2)}{\partial N_2}\right)_{N_2,=\overline{N}_2,E_2,V_2}$$

and we can define another parameter μ,

$$\mu = \left(\frac{\partial \text{ In } \Omega(N,E,V)}{\partial N}\right)_{N=\overline{N},E,V}$$

On comparing with, again recalling thermodynamic formula:

$$\left(\frac{\partial S}{\partial N}\right)_{N,E,V} = -\frac{\mu}{T}$$

We get,

$$\frac{\Delta S}{\Delta(\text{In } \Omega)} = -\frac{\mu}{T} = \text{Constant}$$

Therefore, the equilibrium condition during chemical contact, such that $-\frac{\mu}{T} = k_B$ or $\mu = -k_B T$ becomes:

$$\frac{\mu_1}{T_1} = \frac{\mu_2}{T_2}$$

i.e. $T_1 = T_2$ and $\mu_1 = \mu_2$.

If we combine the three case discussed above such that if we allow exchange of energy, change of volume and exchange of particles simultaneously, the conditions of equilibrium are:

$$T_1 = T_2, P_1 = P_2 \text{ and } \mu_1 = \mu_2.$$

Deriving Thermodynamics from Statistical Route

Now we are ready to derive thermodynamics from statistical considerations. Recipe is as given below:

- For the macrostate (N, V, E) of a given system count the number of all possible microstates accessible to the macrostate denoted by Ω(N, V, E).
- The entropy of the system can be calculated by equation:

  $S = k_B \text{ In } \Omega.$
- The intensive parameters, T, P and μ can then be calculated by the following thermodynamic relations:

$$\frac{1}{T} = \left(\frac{\partial S}{\partial E}\right)_{N,V}$$

$$\frac{P}{T} = \left(\frac{\partial S}{\partial V}\right)_{N,E}$$

$$\frac{\mu}{T} = -\left(\frac{\partial S}{\partial N}\right)_{V,E}$$

Or alternatively,

$$P = \frac{\left(\frac{\partial S}{\partial V}\right)_{N,E}}{\left(\frac{\partial S}{\partial E}\right)_{N,V}} = -\left(\frac{\partial E}{\partial V}\right)_{N,S}$$

and,

$$\mu = -\frac{\left(\frac{\partial S}{\partial N}\right)_{V,E}}{\left(\frac{\partial S}{\partial E}\right)_{N,V}} = \left(\frac{\partial E}{\partial N}\right)_{V,S}.$$

With $U \equiv E$, rest of the thermo dynamics follows from:

- Helmholtz free energy: $F = E - Ts$.
- Gibbs free energy: $G = E - TS + PV$.
- Enthalpy: $H = E + PV = G + TS$.

Furthermore, specific heat at constant volume, C_V and Specific heat at constant pressure, C_P are given by,

$$C_V = \left(\frac{\partial E}{\partial T}\right)_{N,V}$$

$$C_P = \left(\frac{\partial H}{\partial T}\right)_{N,P}$$

In nut shell, once one knows the entropy S(N, E, V) of the system, equilibrium behavior of the system can be known by using thermodynamics.

References

- Thermodynamics: stanford.edu, Retrieved 16, June 2020

- Three-iconic-equations-in-thermodynamics, Statistical-thermodynamics: wikiversity.org, Retrieved 08, March 2020
- Sadi-Carnot-on-Carnot's-theorem-243492415: researchgate.net, Retrieved 13, July 2020
- Equilibrium-physics, science: britannica.com, Retrieved 27, May 2020
- Thermodynamic-potentials: uni-frankfurt.de, Retrieved 20, February 2020

3

Classical and Quantum Ensemble Theory

Classical ensemble theory is used to provide a probability distribution for the state of the system. A few of its major areas of study within this theory are phase space and its properties, micro-canonical ensemble, canonical ensemble, grand canonical ensemble and Liouville's theorem. This chapter discusses in detail these aspects of classical ensemble theory.

Phase Space and its Properties

In the microscopic description of a macroscopic system, made up of a large number of particles ($\sim 10^{23}$), state of the system at any instant t requires specification of instantaneous position and momentum of each of the constituting particles making the system. This leads us to so called work bench of statistical mechanics, a kind of a conceptual space, where all the action in statistical mechanics happens for its statistical study. Phase space provides us the methodology for counting the number of possible states (microstates) for a given system of interest.

Degrees of Freedom

In classical mechanics degrees of freedom, f, corresponds to a number of independent parameters needed to completely define position and motion of a mechanical system. Consider a system made up of a single particle allowed to move in all three directions. The state of such a system at any instant can be described by three position co-ordinates or three momentum co-ordinates, amounting to 3 degrees of freedom. If we constrain this particle to move in a plane, only two position co-ordinates or two momentum co-ordinates are needed to describe its state, reducing the number of degrees of freedom to 2. If we further constrain to move the particle along a straight line, one position or one momentum coordinate are needed to describe its state, thereby system having 1 degrees of freedom.

If a system has N particles, for its complete description we require 3N position co-ordinates or 3N momentum co-ordinates a total of 6N co-ordinates, i.e. f = 3N.

At any given instant, knowledge of degrees of freedom provides us a geometrical way of visualizing state of a mechanical system by means of a space which must have the dimensionality equal to the degrees of freedom of the system. Such a geometrical construct is called a phase space. Each point of the phase space describing a state corresponds to a set of co-ordinates in 6N dimensions $\left(q_1, q_2\, q_3 \dots\dots\dots\dots\, q_{3N};\, p, p_2 p_3 \dots\dots\dots\dots\, p_{3N}\right)$ is called a phase point. The number of position and momentum co-ordinates is two times the number of degrees of freedom for the system. As the system evolves with time, state of the system traverses a trajectory in phase space, going from one state of a system to another and so on.

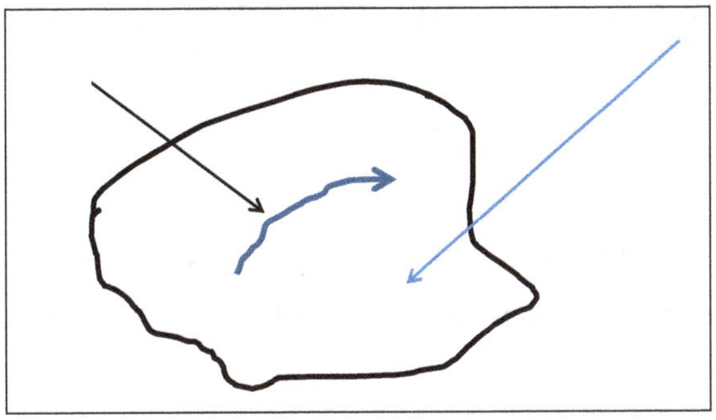

This evolution of phase trajectory with time is determined by the Hamilton's equations of motion,

$$\dot{q}_i = \frac{\partial H(q_i, p_i)}{\partial p_i}$$

$$\dot{p}_i = \frac{\partial H(q_i, p_i)}{\partial q_i}$$

Where, $H(q_i, p_i)$ is the Hamiltonian of the system, which has no explicit dependence on time. Then, H=E is the energy of the system. For a free particle system this amounts to satisfying the condition,

$$\sum_{k=1}^{3N} \frac{p_k^2}{2m} = E$$

In general for a interacting system this amounts to,

$$\sum_{k=1}^{3N} \left(\frac{p_k^2}{2m} + \frac{1}{2} \sum_{j=1}^{3N} V(q_{jk}) \right) = E$$

From here it is obvious that phase point shall traverse a trajectory on the surface of (3N-1) dimensions satisfying the above condition. This space with 6N dimensionality is also referred to as hyperspace of statistical mechanics with which we have to deal with while counting the all possible microstates which a typical system can visit., indeed a very complex entity with N as large as Avogadro's number.

Visualisation of Phase Space

For visualization of phase space, we must try to identify the region of interest, which decides the volume of the phase space available to the system, which as we shall see later ultimately decides the calculation of entropy of the system through an interesting counting procedure. We will at the moment focus only on visualization of the phase space of some simple systems.

Phase Space of a Free Particle Confined to Move Along a Line

Free particle confined to move along a straight line, say x-axis, of length L with constant Energy E.

The Energy of this particle as it moves over this line is the kinetic energy which it possesses, such that we can write:

$$\frac{p^2}{2m} = E$$

This implies that for all values of q on the straight line of length L, momentum can have only two values: $p = \pm\sqrt{2mE}$. This looks like two straight lines parallel to q axis.

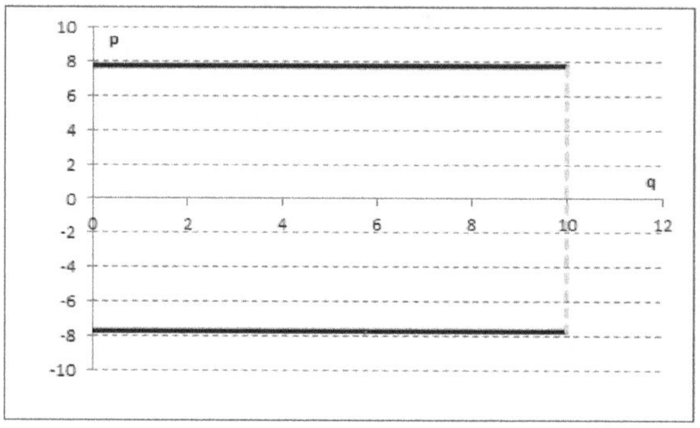

Phase space of a free particle confined to move on a straight line.

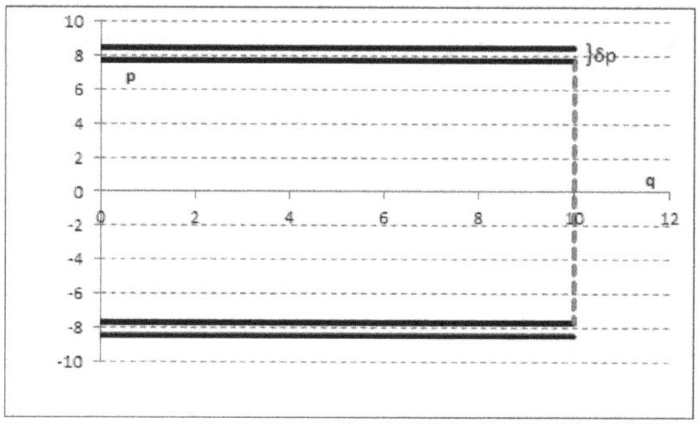

Motion of a particle in one dimension confined to move on a straight line of length L.

Figure above shows the phase space of a free particle moving with constant energy, which are two lines for two possible values of momentum $\left(p = \pm\sqrt{2mE}\right)$ which the particle can take. Figure above shows the two regions of phase space accessible to the free particle between the two parallel lines. Where length L in which particle is confined is a line of length L=10.

Phase Space of a Simple Harmonic Oscillator

The Energy E of a one dimensional harmonic Oscillator,

$$E = \frac{p^2}{2m} + \frac{kq^2}{2}$$

Where, p is linear momentum, q is position co-ordinate, m is the mass of the body executing simple harmonic oscillation in one dimension and k is the force constant. The parametric form of this equation is given by,

$$p = a \sin \omega t, \; q = b \cos \omega t$$

Where $a=\sqrt{2mE}, b=\sqrt{\frac{2E}{k}}$ and $\omega=\sqrt{2mK}$.

Following figure shows the phase space of a linear simple harmonic oscillator whose energy lies between E and E + δE.

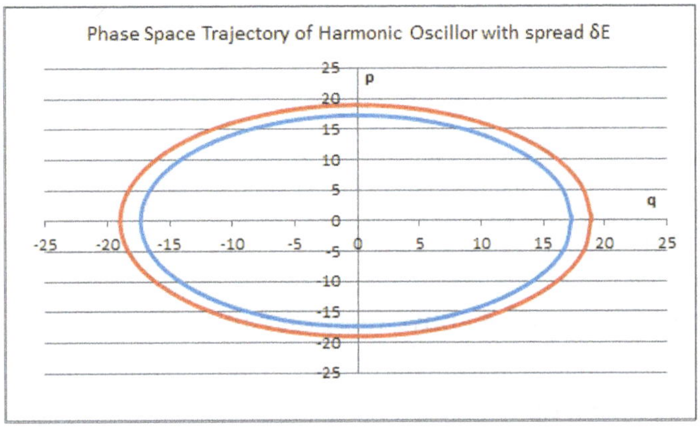

Phase space trajectory of a particle of mass m executing simple harmonic oscillator with energy lying between E and E + δE.

Region lying between the two ellipses is the phase space region accessible to the particle executing simple harmonic oscillations with energy lying between E and E + δE.

Properties of Hyperspace

It is worth exploring some of the geometrical properties of this hyperspace, which we shall find useful at later stage, while calculating the phase space volume and counting the number of possible states.

Distance of a Point in N Dimensional Hyperspace and Consequence of Small Change in Position

Let there be a point $x=(x_1, x_1, \ldots \ldots \ldots \ldots \ldots, x_N)$. Then the distance of this point from the origin is defined as,

$$|x|=\left[\sum_{i=1}^{N} x_i^2\right]^{\frac{1}{2}}$$

Suppose N is a very large number, say Avogadro's number. If some of x_i are not zero then $|x| \sim N^{\frac{1}{2}}$, which is indeed a large number. Furthermore, if each x_i changes by a small amount, $\delta x_i \sim \in$, then $|\delta x| \sim \in N^{\frac{1}{2}}$. Interestingly even if $\in$ is small, $|\delta x|$ can be large, since N is very large.

Volume of a Sphere in N Dimensional Space

Let us note that volume of a sphere in 3-dimensional space is given by $V=\frac{4}{3}\pi r^3$ and surface area which is derivative of V with respect to r is given by $4\pi r^2$. Obviously, therefore, volume (V_n) of an n dimensional sphere of radius r shall be proportional to r^n and surface area (s_n) of this sphere is proportional to nr^{n-1}.

Thus $V_n = C_n r^n$, where C_n is the prortionality constant and can be calculated easily using the value of a Gaussian integral. We know that,

$$\int_{-\infty}^{\infty} dx\, e^{-x^2} \sqrt{\pi}$$

this implies that,

$$\int_{-\infty}^{\infty} dx_1 \dots \dots \int_{-\infty}^{\infty} dx_n \sum_{i=1}^{N} e^{-x_i^2} = \prod_{i}^{N} \int_{-\infty}^{\infty} dx_i e^{-x_i^2} = \pi^{\frac{n}{2}}$$

also, in spherical polar co-ordinates the above integral can be written as,

$$\int_{o}^{\infty} s_n dr\, e^{-r^2} = \int_{o}^{\infty} nC_n r^{n-1} e^{-r^2} dr = \pi^{\frac{n}{2}}$$

therefore,

$$nC_n \int_{o}^{\infty} r^{n-1} e^{-r^2} dr = \pi^{\frac{n}{2}}.$$

Making a change of variable $r^2 = y$,

$$\frac{nC_n}{2} \int_{o}^{\infty} y^{\frac{n}{2}-1} e^{-y} dy = \pi^{\frac{n}{2}}$$

$$\frac{nC_n}{2} \Gamma\left(\frac{n}{2}\right) = \pi^{\frac{n}{2}}$$

Where,

$$\int_{o}^{\infty} y^{\frac{n}{2}-1} e^{-y} dy = \Gamma\left(\frac{n}{2}\right) \text{ is the standard gamma function.}$$

Therefore,

$$C_n = \frac{\pi\frac{n}{2}}{\frac{n}{2}\Gamma\left(\frac{n}{2}\right)} = \frac{\pi^{\frac{n}{2}}}{\Gamma\left(\frac{n}{2}+1\right)}$$

Hence,

$$V_n = \frac{\pi^{\frac{n}{2}}}{\Gamma\left(\frac{n}{2}+1\right)} r^n = \frac{\pi^{\frac{n}{2}}}{\frac{n}{2}!} r^n$$

and the surface area of hypersphere is which is derivative of V_n with respect to r is:

$$S_n = \frac{n\pi^{\frac{n}{2}}}{\Gamma\left(\frac{n}{2}+1\right)} r^{n-1} = \frac{2\pi^{\frac{n}{2}}}{\left(\frac{n}{2}-1\right)!} r^{n-1}$$

Phase Space and Quantum Mechanics

We have seen that classically phase space is a continuum in hyperspace of 3N position c-ordinates and 3N momentum co-ordinates i.e. a total of 6N co-ordinates or degrees of freedom. Each point in the phase space or each set of points $(q_1, q_2 q_3 \ldots \ldots \ldots \ldots q_{3N}; p, p_2 p_3 \ldots \ldots \ldots \ldots p_{3N})$ represents the state of a given macroscopic system. And since q's and p's form a continuum, therefore there are infinitely accessible phase points in phase space.

However, because of Heisenberg's uncertainty principle, quantum mechanics, which states that a particle cannot occupy a certain volume in co-ordinate space and a certain volume in momentum space simultaneously. This means that there is a minimum volume in phase space within which it is not possible to indicate certainly the state of the system. Such that the product of $\Delta p \Delta q$ is of the order of h i.e.

$$\Delta q_x \Delta p_x \geq h,\ \Delta q_y \Delta p_y \geq h,\ \Delta q_z\ \Delta p_z \geq h$$

In general this amounts to for a system of 1 particle a minimum volume of phase space equal to h^3. If there are N particles in the system, it shall amount to ^{3N} volume of phase space. This brings in quantization of the phase space, which has immediate consequence on the counting on probable states for a given state of a macroscopic system.

Microcanonical Ensemble

The highly constrained ensemble the so called micro canonical ensemble, representing a system having constant energy E, constant number of particles N and constant volume V. It describes a system which is completely isolated system from its surroundings, a highly idealized situation seldom encountered in real situations. Within these constraints particles of the system can be in any one of the allowed microstates with equal apriori probability. An effort is made to count the allowed total number of microstates which helps define the probability of each state followed by calculation of entropy which can then be used to calculate the thermodynamic properties of the system under study. But before embarking on this plan for microcanonical ensemble we shall underline the importance of Planck's constant in the light of Heisenberg Uncertainty Principle in counting the microstates in the allowed phase space of the system under study.

Phase Space and Number of Microscopic States

As yet, till this point we have described the enumeration of microstates in terms of calculating the volume of the allowed phase space. This enumeration of microstates is possible through Heisenberg's Uncertainty Principle according to which phase space can be regarded as made up of elementary cells of the minimum volume h^d where d is the dimension of the phase space representing single particle quantum state. In the case of an N particle system with phase space having 6N dimensions, d = 3N.

Microcanonical Ensemble Distribution Function

To define microcanonical distribution function we must ponder on a crucial question, can we have an absolutely isolated system, because in actual systems, there is always some contact with the surroundings and system does not have a sharp value and it will have a width Δ around E. Hence system has a range of energy values $\left(E-\frac{1}{2}\Delta, E+\frac{1}{2}\Delta\right)$, where Δ is small compared to E. Doing this implies that a system has a large number of possible microstates in the phase space lying anywhere between a hyper-shell are equally apriori probable such that,

$$E-\frac{1}{2}\Delta \le H \le E+\frac{1}{2}\Delta$$

This defines a region in phase space having volume,

$$\omega = \int \Theta\left(E+\frac{1}{2}\Delta - H\right)\Theta\left(H-E+\frac{1}{2}\Delta\right) d^{3N}q d^{3N}p$$

Where Θ represents a unit step function defined by,

$$\Theta(x) = \begin{cases} 1 & \text{if } x \ge 0 \\ 0 & \text{otherwise} \end{cases}$$

The step functions sitting as integrand restrict the region of integration defined by equation $E-\frac{1}{2}\Delta \le H \le E+\frac{1}{2}\Delta$. It is obvious that allowed phase space volume is a function of E, N, V and Δ. Then a microcanonical ensemble is a collection of systems, for which probability distribution function ρ(q, p) in equilibrium is given by,

$$\rho(q,p) = \begin{cases} \text{Constant for } E-\frac{1}{2}\Delta \le H \le E+\frac{1}{2}\Delta \\ 0 \quad \text{otherwise} \end{cases}$$

Now if f(q, p) is a physical quantity then the microcanonical ensemble average is given by,

$$\langle f \rangle = \frac{\int \rho(q,p) f(p,q) d^{3N}q\, d^{3N}p}{\int \rho(q,p) d^{3N}q\, d^{3N}p} = \frac{\int f(p,q) d^{3N}q\, d^{3N}p}{\int d^{3N}q\, d^{3N}p}$$

Now we will look at some prototype applications of microcanonical ensemble.

A Two Level System

Let us begin with a simple two level system, i.e. a system which has only two states accessible, consisting of N non-interacting particles in figure below. The lower level is the ground state of the system with zero energy and the other level is the first ecited state of the system with energy ϵ. Both the states are non-degenerate. The higher energy states are not available to the system, because they exceed the total energy available to the system. Let N_0 and N_1 be the number of particles in the ground state and first excited state respectively, such that $N = N_0 + N_1$.

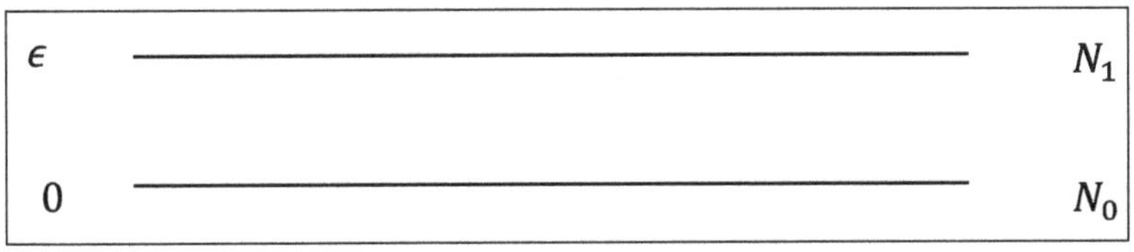

A two non-degenerate level system.

The energy of the system is $E = N_1 \in$. If we assume particles to be distinguishable then the number of microstates is the number of possible ways of choosing N_1 particles out of total N particles to fill the excited level which we call Ω and is given by:

$$\Omega = \binom{N}{N_1} = \frac{N!}{(N-N_1)!N_1!}$$

we are dealing with a thermodynamic system with large N and N_1, we can apply stirling's approximation to calculate entropy S at a system at equilibrium using Boltzmann relation $S = k_B \text{ In } \Omega$:

$$S = k_B\left[N \text{ In } N - (N-N_1)\text{In }(N-N_1) - N_1 \text{ In } N_1\right]$$

Noting that $N_1 = \frac{E}{\in}$, equation above becomes,

$$S = k_B\left[N \text{ In } N - \left(N - \frac{E}{\in}\right)\text{In}\left(N - \frac{E}{\in}\right) - \frac{E}{\in}\text{ In }\frac{E}{\in}\right]$$

which can be simplified further as,

$$S = k_B\left[\left(\frac{E}{\in} - N\right)\text{In}\left(1 - \frac{E}{N\in}\right) - \frac{E}{\in}\text{ In }\frac{E}{N\in}\right]$$

Equation above provides us entropy for the system as a function of extensive variables E and N. Now we are ready to all other measurable thermodynamic quantities like temperature T, energy E and specific heat C_V.

Temperature can be obtained by using the relation $\frac{1}{T} = \left(\frac{\partial S}{\partial E}\right)_{N,V}$

$$\frac{1}{T} = \frac{k_B}{\in}\left[\text{In}\left(1 - \frac{E}{N_\in}\right) + (E - N\in)\left(\frac{N\in}{N\in - E}\right)\left(-\frac{1}{N\in}\right) - \text{In }\frac{E}{N\in} - E\frac{N\in}{E}\frac{1}{N\in}\right]$$

or,

$$\frac{1}{T}=\frac{k_B}{\in}\left[\text{In}\left(1-\frac{E}{N\in}\right)-\text{In}\frac{E}{N\in}\right]$$

or,

$$\frac{1}{T}=\frac{k_B}{\in}\left[\text{In}\left(\frac{N\in}{E}-1\right)\right].$$

Equation above also then gives expression for energy E as,

$$E=\frac{N\in}{1+e^{\frac{\in}{k_BT}}}$$

This effectively means,

$$N_1=\frac{N}{1+e\frac{\in}{e^{k_BT}}}$$

The molar specific heat capacity, when N=Avogadro's number can be obtained by recalling that $C_V=\left(\frac{\partial E}{\partial T}\right)_{N,V}$ which gives,

$$C_V=k_BN\left(\frac{\in}{k_BT}\right)^2\frac{e^{\frac{\in}{k_BT}}}{\left(e^{\frac{\in}{k_BT}}+1\right)^2}$$

Let us look at some limiting cases:

Case: $\frac{\in}{k_BT}\gg 1(\text{i.e. } T\to 0)$

According to equation $N_1=\frac{N}{1+e\frac{\in}{e^{k_BT}}}$, then denominator $\left(\frac{\in}{e^{k_BT}}+1\right)$ is very large and $N_1\to 0$. This means that all particles are sitting in the ground state, total energy in equation $E=\frac{N\in}{1+e^{\frac{\in}{k_BT}}}$, $E=0$ and $\Omega=1$ and hence $S=0$.

Case II: $\frac{\in}{e^{k_BT}}\ll 1$ (i.e. $T\to\infty$)

According to equation $N_1=\frac{N}{1+e\frac{\in}{e^{k_BT}}}$, under this limit since $e^{\frac{\in}{k_BT}}\to 1$, denominator becomes equal to 2 and hence $N_1 = \frac{N}{2}$ i.e. half the particles are in first excited state and half in ground state, energy of the two level system is $E = \frac{N\in}{2}$ and hence entropy $S = k_B N \text{ In } 2$.

Let us look at the behavior of molar specific heat C_V, in both the limits according to equation $C_V = k_B N\left(\frac{\in}{k_B T}\right)^2 \frac{e^{\frac{\in}{k_B T}}}{\left(e^{\frac{\in}{k_B T}}+1\right)^2}$ C_V approaches zero and has maxima at $T = 0.42\frac{\in}{k_B}$ (or $\frac{\in}{k_B} = 2.38$) as shown in figure below:

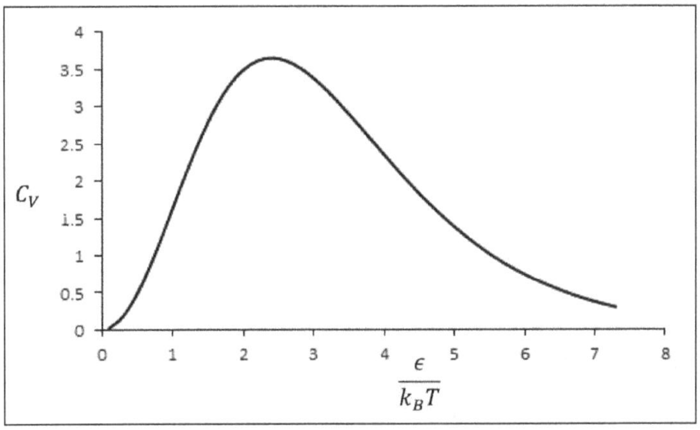

Molar specific heat capacity of a two level system (Schottky Anomaly).

Schottky anomaly found here is a characteristic feature of the systems having a pair of low lying energy levels with other levels of the system lying at a much higher level.

Classical Ideal Gas

Accordingly the volume of the phase space accessible to the phase points in the micro canonical ensemble is ω:

$$\omega = \iint' d^{3N} q \, d^{3N} p$$

Here prime on the integral sign indicates that the particles are fixed in a volume V and the total energy of the system, which in the case of ideal gas is kinetic energy, lies between the limits $E - \frac{1}{2}\Delta \le \frac{p_i^2}{2m} \le E + \frac{1}{2}\Delta$. Furthermore, since Hamiltonian is a function of p_i only the integration over the position co-ordinates can be performed yielding a factor of V^N. The remaining integral is the volume of a 3N dimensional shell enclosed by two hyperspheres of radii $\sqrt{\left[2m\left(E - \frac{1}{2}\Delta\right)\right]}$ and $\sqrt{\left[2m\left(E + \frac{1}{2}\Delta\right)\right]}$. For $\Delta \ll E$, it is thickness of the shell, $\Delta\left(\frac{m}{2E}\right)^{\frac{1}{2}}$, times surface area of the 3N dimensional hypersphere of radius $\sqrt{2mE}$. We had derived it for an n dimensional hypersphere and is,

$$S_n = \frac{2\pi^{\frac{n}{2}}}{\left(\frac{n}{2}-1\right)!} r^{n-1}.$$

Thus volume of the desired 3N dimensional hyper-shell is:

$$\Delta\left(\frac{m}{2E}\right)^{\frac{1}{2}} \frac{2\pi\frac{3N}{2}}{\left(\frac{3n}{2}-1\right)!}(2mE)^{\frac{3N-1}{2}}$$

Therefore, the phase space volume accessible to the representative points of the system is:

$$\omega \cong \frac{\Delta}{E} V^N \frac{(2\pi mE)^{\frac{3N}{2}}}{\left(\frac{3n}{2}-1\right)!}.$$

Hence the number of microstates is:

$$\Omega = \frac{\omega}{h^{3N}} = \frac{3N}{2}\frac{\Delta}{E} \frac{\left(\frac{2\pi mEV^{\frac{2}{3}}}{h^2}\right)^{\frac{3N}{2}}}{\left(\frac{3N}{2}\right)!}.$$

Therefore, entropy is given by using S = k In Ω:

$$S = k_B N\left[\text{In}\left\{V\left(\frac{2\pi mE}{h^2}\right)^{\frac{3}{2}}\right\} - \frac{3}{2}\text{In}\left(\frac{3N}{2}\right) + \frac{3}{2} + \frac{1}{N}\text{In}\left(\frac{3N}{2}\frac{\Delta}{E}\right)\right].$$

For large N, the last term in the expression is negligible hence,

$$S = k_B N\left[\text{In}\left\{V\left(\frac{2\pi mE}{h^2}\right)^{\frac{3}{2}}\right\} - \frac{3}{2}\text{In}\left(\frac{3N}{2}\right) + \frac{3}{2}\right]$$

This expression though expresses S as a function of E, V, N but is still not correct and needs to be corrected for indistinguishability, which led to Gibbs paradox. Once the expression is corrected by dividing Ω by N!.

A Collection of Linear Harmonic Oscillators

Here we are interested in a system of N harmonic oscillators of same ferequency ω in microcanonical ensemble with total energy E. The Hamiltonian of the system is given by,

$$H=\sum_{i=1}^{N}\left(\frac{p_i^2}{2m}+\frac{1}{2}m\,\omega^2 q_i^2\right)$$

We shall look at this problem as a collection of classical oscillators and a collection of quantum mechanical oscillators and compare the two results later.

Classical Approach

Classically it amounts to finding the volume of the phase space ω with in the hypershell of radii lying in the interval $E-\Delta\le\sum_{i=1}^{N}\left(\frac{p_i^2}{2m}+\frac{1}{2}m\omega^2 q_i^2\right)\le E$. This is $\omega=\Delta$ times the surface area of the volume of the phase space bounded by the surface $\sum_{i=1}^{N}\left(\frac{p_i^2}{2mE}+\frac{q_i^2}{\frac{E}{\frac{1}{2}mw^2}}\right)=1$. The volume of this 2N dimensional hypersphere can be obtained from the volume of a generalized sphere in n dimensional space bounded by the surface $\sum_{i=1}^{d}\frac{x_i^2}{a_i^2}=1$ given by,

$$V_d=\frac{\pi^{\frac{d}{2}}}{\frac{d}{2}!}\prod_{i=}^{d}a_i.$$

Using equation above we get,

$$V_{2N}=\frac{\pi^N}{N!}(2mE)^{\frac{N}{2}}\left(\frac{2E}{m\omega^2}\right)^{\frac{N}{2}}=\frac{1}{N!}\left(\frac{2\pi E}{\omega}\right)^N$$

Therefore, the surface area can be obtained by taking derivative of above equation with respect to E and one gets,

$$S_{2N}=\frac{dV}{dE}=\frac{1}{N!}N\left(\frac{2\pi}{\omega}\right)^N E^{(N-1)}=\frac{1}{N!}\left(\frac{2\pi E}{\omega}\right)^N\frac{N}{E}$$

So the volume of the hypershell of phase space with allowed representative points is given by,

$$\omega=\frac{1}{N!}\left(\frac{2\pi E}{\omega}\right)^N\frac{N\Delta}{E}.$$

The number of microstates $\Omega = \frac{\omega}{h^N}$ then is given by,

$$\Omega = \frac{1}{N!}\left(\frac{2\pi E}{h\omega}\right)^N \frac{N\Delta}{E}$$

Entropy of this collection of N linear harmonic oscillators is then given by,

$$S = Nk_B\left[\text{In}\left(\frac{2\pi E}{h\omega}\right) - \text{In } N + 1 + \frac{1}{N}\text{In}\frac{N\Delta}{E}\right].$$

The last term in above equation can be neglected since it has a very small contribution compared other terms.

$$S = Nk_B\left[\text{In}\left(\frac{2\pi E}{Nh\omega}\right) + 1\right]$$

Now we are ready to calculate other thermodynamic quantities such as Temperature and Helmholtz free energy.

Temperature is given by,

$$\frac{1}{T} = \left(\frac{\partial S}{\partial E}\right)_{V,N} = Nk_B\left[\frac{Nh\omega}{2\pi E}\frac{2\pi}{Nh\omega}\right] = \frac{Nk_B}{E}$$

also we get,

$$E = Nk_B T.$$

Helmholtz free energy can now be obtained using the relation F = E – TS

$$F = -T\,Nk_B\left[\text{In}\left(\frac{2\pi E}{Nh\omega}\right)\right] = Nk_B T\left[\text{In}\left(\frac{h\omega}{2\pi k_B T}\right)\right]$$

Quantum Approach

Quantum mechanically, each oscillator has an energy eigen value given by well known relation,

$$\epsilon_i = \left(n_i + \frac{1}{2}\right)\hbar\omega$$

where n_i is a an integer ≥ 0. The total energy of the system is then given by,

$$E = \left(n + \frac{N}{2}\right)\hbar\omega$$

Where $n = n_1 + n_2 + n_3 \dots\dots \dots\dots + n_N$ is a non-negative integer.

Now to calculate the number of microstates corresponding to this total energy. Let us note that this amounts to distributing n packets of energy in N boxes, where each box can contain any number of packets including possibility of zero packets. If we wish to create N boxes in a box we require N-1 partitions, so to count the number of possible ways to distribute n energy packets in N boxes is equivalent to arranging a total of n packets and N-1 partitions together, i.e. permuting n+N-1 out of which n packets among them are indistinguishable and N-1 partitions among them are indistinguishable.

Therefore,

$$\Omega = \frac{(n+N-1)!}{n!(N-1)!}$$

The entropy of the system after using stirlings approximation is then given by,

$$S = k_B \text{ In } \Omega = k_B\left[(n+N-1) \text{ In } (n+N-1)\right] \\ -n \text{ In } n-(N-1) \text{ In } (N-1)$$

Since in thermodynamic limit, $N \gg 1,\ N-1 \cong N$ and $n = \left(\frac{E}{\hbar\omega} - \frac{N}{2}\right)$, equation above becomes,

$$S = k_B\left[\left(\frac{E}{\hbar\omega}+\frac{N}{2}\right)\text{In}\left(\frac{E}{\hbar\omega}+\frac{N}{2}\right)-\left(\frac{E}{\hbar\omega}-\frac{N}{2}\right)\text{In}\left(\frac{E}{\hbar\omega}-\frac{N}{2}\right)-N \text{ In } N\right]$$

or,

$$S = Nk_B\left[\left(\frac{E}{N\hbar\omega}+\frac{1}{2}\right)\text{In}\left(\frac{E}{N\hbar\omega}+\frac{1}{2}\right)-\left(\frac{E}{N\hbar\omega}-\frac{1}{2}\right)\text{In}\left(\frac{E}{N\hbar\omega}-\frac{1}{2}\right)\right]$$

Now we are ready to derive all thermodynamic relations,

$$\frac{1}{T} = \left(\frac{\partial S}{\partial E}\right)_{N,V} = \frac{k_B}{\hbar\omega}\left(\text{In}\left(\frac{E}{N\hbar\omega}+\frac{1}{2}\right)-\text{In}\left(\frac{E}{N\hbar\omega}-\frac{1}{2}\right)\right)$$

or,

$$\frac{\left(\frac{E}{N\hbar\omega}+\frac{1}{2}\right)}{\left(\frac{E}{N\hbar\omega}-\frac{1}{2}\right)} = e\frac{\hbar\omega}{e^{k_B T}}$$

or,

$$E = N\ \hbar\omega\left\{\frac{1}{2}+\frac{1}{e\frac{\hbar\omega}{e^{k_B T}}-1}\right\}$$

It is interesting to ask what happens to equation above, a quantum mechanical result in the classical limit i.e. when $\frac{\hbar\omega}{k_B T} \ll 1$.

$$E \approx N\left\{\frac{\hbar\omega}{2} + \hbar\omega\frac{k_B T}{\hbar\omega}\right\} \approx Nk_B T$$

Same result as the classical result ($E = Nk_B T$).

Furthermore, specific heat at constant volume $C_V = \left(\frac{\partial E}{\partial T}\right)_{N,V}$ can be obtained as,

$$C_V = Nk_B\left(\frac{\hbar\omega}{k_B T}\right)^2 \frac{e^{\frac{\hbar\omega}{k_B T}}}{\left(e^{\frac{\hbar\omega}{k_B T}} - 1\right)^2}$$

In the classical limit, it becomes,

$$C_V = Nk_B\left(\frac{\hbar\omega}{k_B T}\right)^2 \frac{1}{\left(\frac{\hbar\omega}{k_B T}\right)^2} = Nk_B$$

Similarly Helmholtz free energy in quantum and classical limits can be obtained.

Figures below show respectively the plot of $\frac{E}{N\text{h}\omega}$ and C_V against $\frac{k_B T}{\text{h}\omega}$.

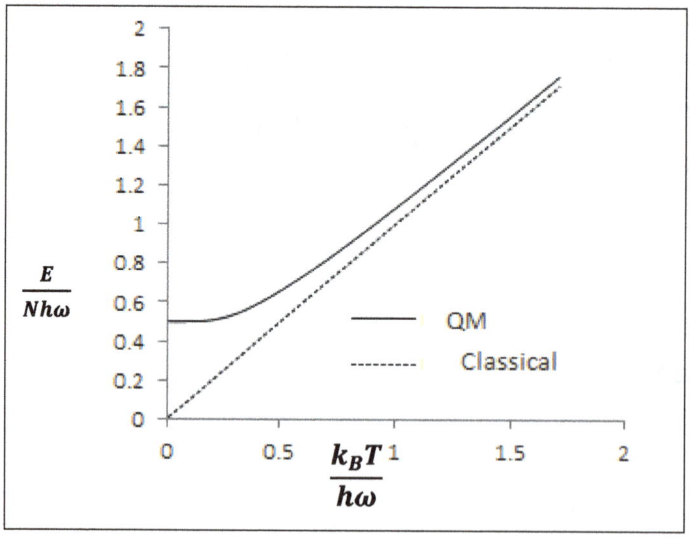

$\frac{E}{N\text{h}\omega}$ versus $\frac{k_B T}{\text{h}\omega}$

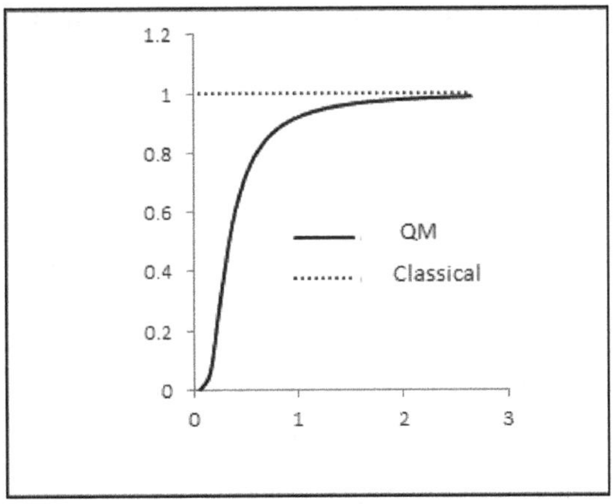

$\frac{C_V}{Nk_B}$ versus $\frac{k_b T}{h\omega}$ for a collection of N Harmonic Oscillators.

This example actually represents model of an Einstein solid made of N uncoupled quantum oscillators of same frequency.

A System of Spin $\frac{1}{2}$ Particles

We now consider an isolated system of N particles of spin $\frac{1}{2}$, i.e. $\vec{\sigma_\iota} = \frac{1}{2}\hbar\hat{z} = \sigma_i^z\hat{z}$.. Each particle in the system, therefore has a magnetic moment $\mu\vec{\sigma}$. Placed in a magnetic field $\vec{H}$ such a particle has only two orientations, spin up (↑) or spin down (↓) with respect to the magnetic field. Let the magnetic field be in the direction of Z-axis.

The Hamiltonian for such a system is given by,

$$\mathcal{H} = \sum_{i=1}^{N} \mu\vec{\sigma_\iota}.\vec{H} = -\mu H \sum_{i=1}^{N} \sigma_i^z$$

The Hamiltonian does not have a term for the motion of the particles except interaction with the magnetic field. System has number of particles N and total magnetic moment M as extensive parameters, Volume has no role to play in this situation.

Suppose that at equilibrium system has N_+ particles with spin up and N_- particles with spin down, such that,

$$N = N_+ + N_-$$

The net magnetic moment of the system is then,

$$M = \mu(N_+ - N_-)$$

Solving ($N = N_+ + N_-$) and ($M = \mu(N_+ - N_-)$) for N_+ and N_-, we have

$$N_\pm = \frac{N}{2}\left(1 \pm \frac{M}{\mu N}\right)$$

So to count the number of microstates it is merely a matter of selecting N_+ number of particles out of N particles and is given by,

$$\Omega(M,N)=\frac{N!}{N_+!\,N_-!}.$$

Entropy can then be obtained from Boltzmann relationship $S=k_B\ \mathrm{In}\ \Omega(M,N)$ as,

$$S(M,N)=k_B\left(\mathrm{In}\ N!-\mathrm{In}\ N_+!-\mathrm{In}\ N_-!\right).$$

In thermodynamic limit using Stirlings Formula and ($N_\pm=\frac{N}{2}\left(1\pm\frac{M}{\mu N}\right)$), we have,

$$S(M,\,N)=k_B\left(N\ \mathrm{In}\ N-\frac{N}{2}\left(1+\frac{M}{\mu N}\right)\mathrm{In}\frac{N}{2}\left(1+\frac{M}{\mu N}\right)-\frac{N}{2}\left(1-\frac{M}{\mu N}\right)\mathrm{In}\frac{N}{2}\left(1-\frac{M}{\mu N}\right)\right)$$

Or,

$$S(M,\,N)=Nk_B\left(\mathrm{In}\ 2-\frac{1}{2}\left(1+\frac{M}{\mu N}\right)\mathrm{In}\left(1+\frac{M}{\mu N}\right)-\frac{1}{2}\left(1-\frac{M}{\mu N}\right)\mathrm{In}\left(1-\frac{M}{\mu N}\right)\right)$$

Equation above is the entropy of the system in terms of extensive variables M and N.

Knowing that $\left(\frac{\partial S}{\partial M}\right)_{E,V,N}=-\frac{H}{T}$

We have,

$$-\frac{H}{T}=\frac{k_B}{2\mu}\mathrm{In}\ \frac{\mu N-M}{\mu N-M}$$

which can be further simplified to give the expression for magnetic moment,

$$M=\mu N\frac{\left(e^{\frac{\mu H}{k_BT}}-e^{-\frac{\mu H}{k_BT}}\right)}{\left(e^{\frac{\mu H}{k_BT}}+e^{-\frac{\mu H}{k_BT}}\right)}=\mu N\ \tan h\frac{\mu H}{k_BT}.$$

This is the equation of state of a system with N particles of spin $\frac{1}{2}$.

Before we end, it will be interesting to explore $\mathrm{In}\ \Omega\ (M,N)$ using

$$\left(S(M,\,N)=Nk_B\left(\mathrm{In}\ 2-\frac{1}{2}\left(1+\frac{M}{\mu N}\right)\mathrm{In}\left(1+\frac{M}{\mu N}\right)-\frac{1}{2}\left(1-\frac{M}{\mu N}\right)\mathrm{In}\left(1-\frac{M}{\mu N}\right)\right)\right)$$

noting that for $0<x<1$, $\mathrm{In}(1\pm x)\cong\pm x-\frac{x^2}{2}\pm\frac{x^3}{3}-\ldots\ \ldots\ \ldots\ \ldots$ we have,

$$\mathrm{In}\ \Omega\ (M,N)=N\left(\mathrm{In}\ 2-\frac{1}{2}\left(\frac{M}{\mu N}\right)^2\right)$$

we have,

$$\Omega(M,N) = e^{(\ln 2)} e^{\left(-\frac{1}{2N}\left(\frac{M}{\mu}\right)^2\right)} \approx e^{\left(-\frac{1}{2N}\left(\frac{M}{\mu}\right)^2\right)}$$

This is a Gaussian distribution with a spread of $\sqrt{N}$ as shown in figure.

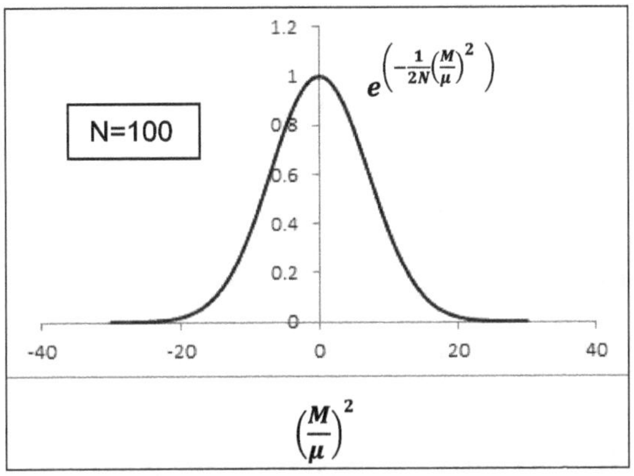

Gaussian distribution equation $\Omega(M,N) = e^{(\ln 2)} e^{\left(-\frac{1}{2N}\left(\frac{M}{\mu}\right)^2\right)} \approx e^{\left(-\frac{1}{2N}\left(\frac{M}{\mu}\right)^2\right)}$.

Canonical Ensemble

System in Contact with a Heat Reservoir

We consider a small system A_1 characterized by E_1, V_1 and N_1 in thermal interaction with a heat reservoir A_2 characterized by E_2, V_2 and N_1 in thermal interaction such that $A_1 \ll A_2$, A_1 has hence fewer degrees of freedom than A_2.

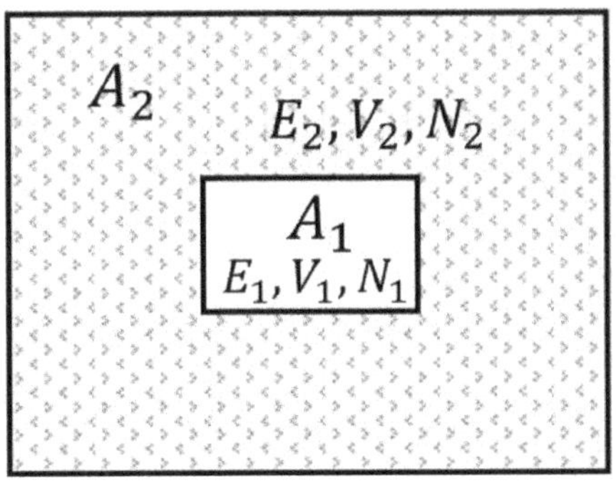

$$E_2 \gg E_1 N_1 = \text{const.}$$

$$N_2 \gg N_1 N_2 = \text{const.}$$

with

$$E_1 + E_2 = E = \text{const.}$$

Both systems are in thermal equilibrium at temperature T. The wall between them allows interchange of heat but not of particles. The system A_1 may be any relatively small macroscopic system such as, for instance, a bottle of water in a lake, while the lake acts as the heat reservoir A_2.

We note that the energy E_1 is not fixed, only the total energy $E = E_1 + E_2$ of the combined system.

Hamilton Function

The Hamilton function of the combined system A is,

$$H(q, p) = H_1(q(1), p(1)) + H_2(q(2), p(2)),$$

where we have used the notation,

$$q = (q(1), q(2)), \quad p = (p(1), p(2)).$$

Microcanonical Ensemble of the Combined System

Since the combined system A is isolated, the distribution function in the combined phase space is given by the microcanonical distribution function $\rho(q, p)$,

$$\rho(q,p) = \frac{\delta(E - H(q, p))}{\int dq\, dq\, \delta(E - H(q, p))}, \quad \int dq\, dq\, \delta(E - H) = \Omega(E),$$

where $\Omega(E)$ is the density of phase space.

Tracing out A_2. It is not the distribution function $\rho(q, p) = \rho(q(1), p(1), q(2), p(2))$ of the total system A that we are interested in, but in the distribution function $\rho_1(q(1), p(1))$ of the small system A_1. One hence needs to trace out A_2:

$$\begin{aligned} \rho_1(q(1),p(1)) &= \int dq\,(2)\, dp\,(20)\, \rho(q(1), q(2), p(2)) \\ &= \frac{\int dq\,(2)\, dp(2)\, \delta(E - H_1 - H_2)}{\Omega(E)} \\ &\equiv \frac{\Omega_2(E - H_1)}{\Omega(E)}. \end{aligned}$$

Where $\Omega_2(E_2) = \Omega(E - H_1)$ is the phase space density of A_2.

Small E_1 Expansion

Now, we make use of the fact that A_1 is a much smaller system than A_2 and therefore the energy E_1 given by H_1 is much smaller than the energy of the combined system:

$$E_1 \ll E.$$

In this case, we can approximate $(\equiv \frac{\Omega_2(E-H_1)}{\Omega(E)})$ by expanding the slowly varying logarithm of $\Omega_2(E_2) = \Omega_2(E-H_1)$ around the $E_2 = E$ as,

$$\text{In } \Omega_2(E_2) \text{ In } \Omega_2(E-H_1) \simeq \text{In } \Omega_2(E) - \left[\frac{\partial \text{ In } \Omega_2}{\partial E_2}\right]_{E_2=E} H_1 + \dots$$

and neglect the higher-order terms since $H_1 = E_1 \ll E$.

Derivatives of the Entropy

$$S = k_B \text{ In}\left(\frac{\Gamma(E,V,N)}{\Gamma_o}\right) = k_B \text{ In}\left(\frac{\Omega(E)\Delta}{\Gamma_o}\right),$$

where Δ is the width of the energy shell, we find that derivatives of the entropy like,

$$\frac{1}{T} = \frac{\partial S}{\partial E} = k_B \frac{\partial \text{In}\Omega(E)}{\partial E}$$

can be taken with respect to the logarithm of the phase space density $\Omega(E)$.

Boltzmann Factor

Using $(\frac{1}{T} = \frac{\partial S}{\partial E} = k_B \frac{\partial \text{In}\Omega(E)}{\partial E})$ for the larger system A_2 we may rewrite $(\text{In } \Omega_2(E_2) \text{ In } \Omega_2(E-H_1) \simeq \text{In } \Omega_2(E) - \left[\frac{\partial \text{ In } \Omega_2}{\partial E_2}\right]_{E_2=E} H_1 + \dots)$ as,

$$\Omega_2(E-H_1) = \exp\left[\text{In } \Omega_2(E) - \frac{\partial \text{ In } \Omega_2(E_2)}{\partial E_2}\bigg|_{E_2=E} H_1 + \dots\right]$$

$$= \Omega(E)\exp\left[-\frac{H_1}{k_B T_2}\right].$$

The temperature T_2 of the heat reservoir A_2 by whatever small amount of energy the large system A_2 gives to the small system A_1. Both systems are thermally coupled, such that $T_1 = T_2 = T$. We hence find with,

$$\rho_1(q(1),p(1)) = \int dq\,(2)\,dp\,(20)\,\rho(q(1), q(2), p(2))$$

$$= \frac{\int dq\,(2)\,dp(2)\,\delta(E-H_1-H_2)}{\Omega(E)}$$

$$\equiv \frac{\Omega_2(E-H_1)}{\Omega(E)}.$$

$$\rho_1(q(1),p(1)) = \frac{\Omega_2(E)}{\Omega(E)} e^{-\frac{H_1}{k_B T}} \propto e^{-\frac{H_1}{k_B T}}.$$

The factor exp[$-H_1$/(kBT)] is called the Boltzmann factor.

Distribution Function of the Canonical Ensemble

The prefactor $\Omega_2(E)/\Omega(E)$ in ($\rho_1(q(1),p(1)) = \frac{\Omega_2(E)}{\Omega(E)} e^{-\frac{H_1}{k_B T}} \propto e^{-\frac{H_1}{k_B T}}$) is independent of H_1. We may hence obtain the the normalization of ρ_1 alternatively by integrating over the phase space of A_1:

$$\rho(q(1),p(1)) = \frac{e-\beta H_1(q(1),p(1))}{\int dq(1)dp\,(1)e^{-\beta H_1(q(1),p(1))}},\ \beta = \frac{1}{k_B T}.$$

Boltzmann Factor

The probability P_α of finding the system A_1 (which is in thermal equilibrium with the heat reservoir A_2) in a microstate α with energy E_α is given by,

$$P_\alpha = \frac{e^{-\beta E\alpha}}{\sum_\alpha e^{-\beta E\alpha}} \quad \text{Boltzmann distribution}$$

when rewriting ($\rho(q(1),p(1)) = \frac{e-\beta H_1(q(1),p(1))}{\int dq(1)dp\,(1)e^{-\beta H_1(q(1),p(1))}},\ \beta = \frac{1}{k_B T}$) in terms of P_α.

- The number of states $\Omega_2(E_2) = \Omega_2(E - H_1)$ accessible to the reservoir is a rapidly increasing function of its energy.
- The number of states $\Omega_2(E_2) = \Omega_2(E - H_1)$ accessible to the reservoir decreases therefore rapidly with increasing $E_1 = E - E_2$. The probability of finding states with large E_1 is accordingly also rapidly decreasing.

The exponential dependence of P_α on E_α in previous equation expresses this fact in mathematical terms.

Example: Suppose a certain number of states accessible to A_1 and A_2 for various values of their respective energies, as given in the figure, and that the total energy of the combined system is 1007.

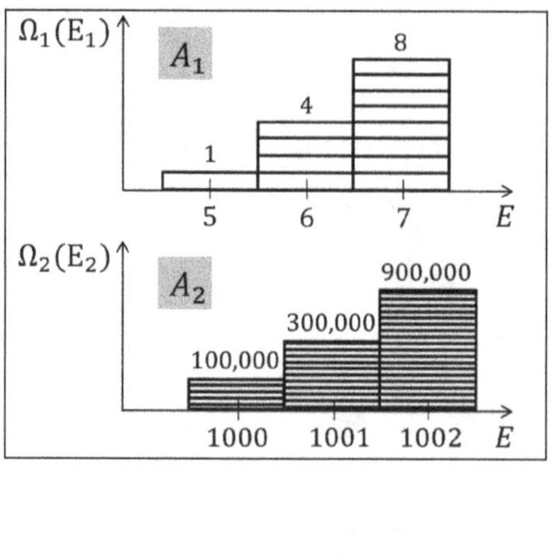

- Let A_1 be in a state α with energy 6. E_2 is then in one of the $3 \cdot 10^5$ states with energy 1001.
- If A_1 is in a state γ with energy 7, the reservoir must be in one of the $1 \cdot 10^5$ states with energy 1000.

The number of realizations of states with $E_1 = 6$ the ensemble contains is hence much higher than the number of realization of state with $E_1 = 7$.

1. Canonical ensemble: An ensemble in contact with a heat reservoir at temperature T is called a canonical ensemble, with the Boltzmann factor $\exp(-\beta E_\alpha)$ describing the canonical distribution ($P_\alpha = \frac{e^{-\beta E\alpha}}{\sum_\alpha e^{-\beta E\alpha}}$).

2. Energy distribution function: The Boltzmann distribution ($P_\alpha = \frac{e^{-\beta E\alpha}}{\sum_\alpha e^{-\beta E\alpha}}$) provides the probability P_α to find an individual microstates α. There are in general many microstates in a given energy, for which,

$$P(E) \sum_{E<E\alpha<E+\Delta} P\alpha \propto \Omega(E)e^{-\beta E},$$

is the corresponding energy distribution function. $\Omega(E) = \Omega_1(E)$ is, as usual, the density of phase space.

- P(E) is rapidly decreasing for increasing energies due to the Boltzmann factor $\exp(-\beta E_\alpha)$.
- P(E) is rapidly decreasing for decreasing energies due to the decreasing phase space density Ω(E).

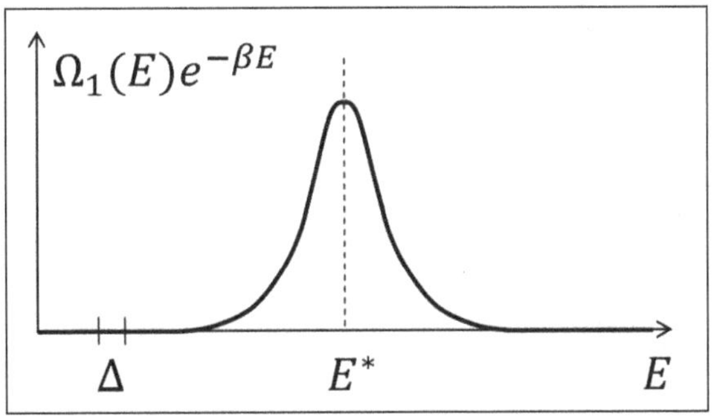

The energy density is therefore sharply peaked.

Canonical Partition Function

We rewrite the distribution function ($\rho(q(1),p(1)) = \frac{e-\beta H_1(q(1),p(1))}{\int dq(1)dp\,(1)e^{-\beta H_1(q(1),p(1))}}$, $\beta = \frac{1}{k_B T}$) of the canonical ensemble as,

$$\rho(q, p) == \frac{e^{-\beta H(q,p)}}{\int d^{3N}q\,d^{3N}p\,e^{-\beta H(q,p)}},$$

where we dropped all the indices "1" for simplicity, though in fact we are still describing the properties of a "small" system (which is nevertheless macroscopically big) in thermal equilibrium with a heat reservoir.

- Partition function: The canonical partition function ("kanonische Zustandssumme") Z_N is defined as:

$$Z_N = \int \frac{d^{3N}q\, d^{3N}}{h^{3N}N!} e^{-\beta H(q,p)}.$$

It is proportional to the canonical distribution function ρ(q, p), but with a different normalization, and analogous to the microcanonical space volume Γ(E) in units of Γ_o:

$$\frac{\Gamma(E)}{\Gamma_o} = \frac{1}{h^{3N}N!}\int_{E<H(q,p)<E+\Delta} d^{3N}q\, d^{3N}p$$
$$= \int \frac{d^{3N}q\, d^{3N}}{h^{3N}N!}\left(\Theta(E+\Delta-H)-\Theta(E-H)\right),$$

where Θ is the step function.

- Free energy: We will show that it is possible to obtain all thermodynamic observables by differentiating the partition function Z_N. We will prove in particular that,

$$F(T,V,N) = -k_B T \,\mathrm{In}\, Z_N(T), \quad Z_N = e^{-\beta F(T,V,N)}$$

where F(T, V, N) is the Helmholtz free energy.

Proof: In order to proof above equation we perform the differentiation,

$$\frac{\partial}{\partial\beta}\mathrm{In}\, Z_N = \frac{1}{Z_N}\frac{\partial Z_N}{\partial\beta}$$
$$= \left[\frac{\partial}{\partial\beta}\int\frac{dq\,dp}{h^{3N}N!}e^{-\beta H}\right]\Big/\left[\int\frac{dq\,dp}{h^{3N}N!}e^{-\beta H}\right]$$
$$= \frac{\int dq\,dp(-H)e^{-\beta H}}{\int dq\,dp\, e^{-\beta H}}$$
$$= -\langle H\rangle = -U.$$

where we have used the shortcut $dq\,dp = d^{3N}q d^{3N}p$ and that $\langle H\rangle = E = U$ is the internal energy. $U = \partial(\beta F)/\partial\beta$, we find that,

$$-\frac{\partial}{\partial\beta}\mathrm{In}\, Z_N = U = \frac{\partial}{\partial\beta}(\beta F), \ \mathrm{In}\, Z_N = -\beta F, \ Z_N = e^{-\beta F},$$

which is what we wanted to prove.

- Integration constant: Above derivation allows to identify ln Z_N = $-\beta F$ only up to an integration constant (or, equivalently, Z_N only up to a multiplicative factor). Setting this constant to zero results in the correct result for the ideal gas.

- Thermodynamic properties: Once the partition function Z_N and the free energy F(T, V, N) = $-k_BT \ln Z_N$ (T, V, N) are calculated, one obtains the pressure P, the entropy S and the chemical potential μ as usual via,

$$P = -\left(\frac{\partial F}{\partial V}\right)_{T,N}, \quad S = -\left(\frac{\partial F}{\partial T}\right)_{V,N}, \quad \mu = \left(\frac{\partial F}{\partial N}\right)_{T,V}.$$

- Specific heat: The specific heat C_V is given in particular by,

$$\frac{C_V}{T} = \left(\frac{\partial S}{\partial T}\right)_V = -\frac{\partial^2 F}{\partial T^2} = \left(\frac{\partial^2}{\partial T^2}\right)(k_B T \operatorname{In} Z_N)$$

where we have used $F = -k_BT \ln Z_N$.

Canonical vs. Microcanonical Ensemble

We have seen that the calculations in the microcanonical and canonical ensembles reduce to a phase space integration and a calculation of a thermodynamic potential:

	Microcanonical ensemble	Canonical ensemble
Phase space integration	Density of states: $\Omega_N(E) = \int d^{3N} q\, d^{3N} p \delta(E - H)$	$Z_n(T) = \int \frac{d^{3N} q\, d^{3N} p}{h^{3N} N!} e^{-\beta H(q,p)}$
Thermodynamic potential	$S(E,V,N) = k_B \operatorname{In}\left(\frac{\Omega_N(E)\Delta}{h^{3N} N!}\right)$	$F(T,V,N) = -k_B T \operatorname{In} Z_N(T)$

- Laplace transforms: The relation between the density of states $\Omega_N(E)$ and the partition function Z_N (T) can be defined as a Laplace transformation in the following way. We use the definition

$$(\rho(q,p) = \frac{\delta(E - H(q,p))}{\int dq\, dq\, \delta(E - H(q,p))}, \quad \int dq\, dq\, \delta(E - H) = \Omega(E),)$$ of the density of states $\Omega(E)$,

$$\int dq\, dp\, \delta(E - H) = \Omega(E), \quad H = H(q,p),$$

in order to obtain,

$$\int_0^\infty \frac{dE\, e^{-\beta E}}{h^{3N} N!} \Omega_N(E) = \int d^{3N}q\, d^{3N}p \int_0^\infty \frac{dE e^{-\beta E}}{h^{3N} N!} \delta(E - H)$$

$$= \int \frac{d^{3N}q\, d^{3N}p}{h^{3N} N!} e^{-\beta H(q,p)} = Z_N(T).$$

We have thus shown that Z_N (T) is the Laplace transform† of Ω_N (E).

- Additive Hamilton functions: In both the microcanonical and in the canonical ensemble we have to perform an integration which is usually difficult. When the Hamilton function is additive, $H = \sum_i H_i$, the integration in the canonical ensemble can be factorized, which is not the case for the microcanonical ensemble. Therefore, it is usually easier to calculate in the canonical ensemble than in the microcanonical ensemble.

Additivity of F(T, V, N)

An important property of the free energy is that it has to be additive.

- Non-interacting systems: Let us consider two systems in thermal equilibrium. Neglecting the interaction among the systems, the total Hamilton function can be written as a sum of the Hamiltonians of the individual systems,

$H = H_1 + H_2$, $N = N_1 + N_2$.

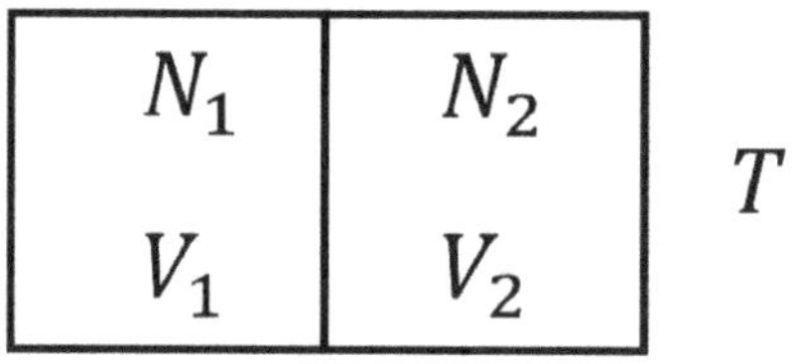

- Multiplication of partition functions: The partition function of the total system is,

$$Z_N(T,V) = \frac{1}{h^{3N} N_1! N_2!} \int d^{3N}q\, d^{3N}p\, e^{-\beta(H_1+H_2)},$$

Where we have made use of the fact that there is not exchange of particles between the two systems. The factor in the denominator is therefore proportional to $N_1!\, N_2!$ and NOT to $N!$. It then follows that the partition function factorizes,

$$\begin{aligned} Z_N(T,V) &= \frac{1}{h^{3N_1} N_1!} \int d^{3N_1}q\, d^{3N_1}p\, e^{-\beta H_1(q_1,p_1)} \\ &\times \frac{1}{h^{3N_2} N_2!} \int d^{3N_2}q\, d^{3N_2}p\, e^{-\beta H_2(q_2,p_2)} \\ &= Z_{N1}(T,V_1) Z_{N2}(T,V_2), \end{aligned}$$

and that the free energy $F = -k_B T \ln Z_N$ is additive:

$F(T, V, N) = F_1(T, V_1, N_1) + F_2(T, V_2, N_2)$.

- Convolution of densities of states: That the overall partition function factorizes follows also from the fact that the density of states $\Omega(E)$ of the combined system,

$$\begin{aligned} \Omega(E) &= \int d^{3N}q\, d^{3N}p\, \delta(E - H_1 - H_2) \\ &= \int d^{3N_1}q\, d^{3N_1}p\, d^{3N_2}p\, d^{3N_2}p \int dE_2 \delta(E - H_1 - E_2)\delta(E_2 - H_2) \\ &= \int dE_2\, \Omega_1(E - E_2)\Omega_2(E_2), \end{aligned}$$

is given by the (convolution) of the density of states $\Omega_i(E_i)$ of the individual systems. Using the representation $(=\int\frac{d^{3N}q\,d^{3N}p}{h^{3N}N!}e^{-\beta H(q,p)}=Z_N(T))$ for the partition function we obtain,

$$Z_N=\int\frac{dE\,e^{-\beta E}}{h^{3N}N_1!N_2}\Omega(E)$$
$$=\int\frac{dE\,e^{-\beta(E_1+E_2)}}{h^{3N}N_1!N_2!}\underbrace{\int dE_2\,\Omega_1(E-E_2)\Omega_2(E_2)}_{\Omega_1(E_1)}.$$

A change of the integration variable from dE to dE_1 then leads again to,

$Z_N(T,V)=Z_{N1}(T,V_1)Z_{N2}(T,V_2)$.

Note that this relation is only valid if $H = H_1 + H_2$ and $H_{12} = 0$.

Ideal Gas in the Canonical Ensemble

We consider now the ideal gas in the canonical ensemble, for which the Hamilton function,

$$H=\sum_{i=1}^{N}\frac{\vec{p}_i^{\,2}}{2m},\quad Z_N(T,V)=\int\frac{d^{3N}q\,d^{3N}p}{h^{3N}N!}e^{-\beta\sum_{i=1}^{N}\vec{p}_i^{\,2}(2m)},$$

contains just the kinetic energy.

1. Factorization: The integral leading to Z_N factorizes in above equation,

$$Z_N(T,V)=\frac{V_N}{N!}\left(\int_{-\infty}^{+\infty}\frac{dp}{h}e^{-\beta\frac{p^2}{2m}}\right)^{3N}$$
$$=\frac{V^N}{N!}\left(\int_{-\infty}^{+\infty}\frac{\sqrt{2k_BTm}}{h}e-x^2dx\right)^{3N},$$

where we have used the variable substitution,

$$x^2=\frac{p^2}{2k_BTm},\qquad dx=\frac{dp}{\sqrt{2k_BTm}},\qquad \int_{-\infty}^{+\infty}dxe^{-x^2}=\sqrt{\pi}.$$

Thermal wavelength. Evaluating $(=\frac{V^N}{N!}\left(\int_{-\infty}^{+\infty}\frac{\sqrt{2k_BTm}}{h}e-x^2dx\right)^{3N})$ explicitly with the help of $(x^2=\frac{p^2}{2k_BTm},\quad dx=\frac{dp}{\sqrt{2k_BTm}},\quad \int_{-\infty}^{+\infty}dxe^{-x^2}=\sqrt{\pi}.)$ we get,

$$Z_N(T,N)=\frac{V^N}{N!}\left(\frac{\sqrt{2\pi mk_BT}}{h}\right)^{3N}\equiv\frac{1}{N!}\left(\frac{V}{\lambda_T^3}\right)^N,$$

where we have defined the thermal wavelength λ_T as,

$$\lambda_T = \frac{h}{\sqrt{2\pi m k_B T}}.$$

For air (actually nitrogen, N_2, with $m = 4.65 \cdot 10^{-26}$ kg) at $T = 298$ K, the thermal wavelength is 0.19 A°, which is actually smaller than the Bohr radius. Quantum mechanical effects start to play a role only once λ_T becomes larger than the typical interparticle separation.

2. Thermal momentum: Heisenberg's uncertainty principle $\Delta x \cdot \Delta p \sim h$ allows to define a thermal momentum p_T as,

$$p_T = \frac{h}{\lambda_T} = \sqrt{2\pi m k_B T}, \quad \frac{p_T^2}{2m} = \pi k_B T = \frac{2\pi}{3} E_{kin}, \quad E_{kin} = \frac{3}{2} k_B T,$$

The thermal momentum P_T is hence of the same order of magnitude as the average momentum $\bar{p}$ of the gas, as defined by $E_{kin} = \bar{p}^2/(2m)$, but not identical.

3. Free energy: From $\left(Z_N(T,N) = \frac{V^N}{N!}\left(\frac{\sqrt{2\pi m k_B T}}{h}\right)^{3N} \equiv \frac{1}{N!}\left(\frac{V}{\lambda_T^3}\right)^N, \right)$ we obtain (with log N! ≈ N log N – N)

$$\begin{aligned} F(T,V,N) &= -k_B T \, \text{In}\left(\frac{1}{N!}\left(\frac{V}{\lambda_T^3}\right)^N\right) \\ &= -k_B T\left(\text{In}\frac{1}{N!} + \text{In}\left(\frac{V}{\lambda_T^3}\right)^N\right) \\ &= -k_B T\left(-N \, \text{In} \, N + N + N \, \text{In}\frac{V}{\lambda_T^3}\right) \end{aligned}$$

and hence,

$$F(T,V,N) = N k_B T \left\{\text{In}\left(\frac{V}{N\lambda_T^3}\right) + 1\right\}$$

for the free energy of the ideal gas.

4. Entropy: Using,

$$\frac{\partial \lambda_T}{\partial T} = \frac{\lambda_T}{2T}, \quad \frac{\partial}{\partial T} \text{In}\left(\frac{V}{N\lambda_T^3}\right) = -3\frac{\partial \, \text{In} \lambda_T}{\partial T} = \frac{3}{2T}$$

we then have,

$$S = -\left(\frac{\partial F}{\partial T}\right)_{V,N} = N k_B \left[\text{In}\left(\frac{V}{N\lambda_T^3}\right) + 1\right] + N k_B T\left[\frac{3}{2T}\right],$$

which results in the Sackur-Tetrode equation,

$$S = Nk_B \left\{ \text{In} \frac{V}{N\lambda_T^3} + \frac{5}{2} \right\}.$$

namely with the microcanonical Sackur-Tetrode equation,

$$S = k_B N \left\{ \text{In} \left[\left(\frac{(4\pi mE)}{3h^2 N} \right)^{3/2} \frac{V}{N} \right] + \frac{5}{2} \right\}$$

one finds that they coincide when $E/N = 3k_B T/2$.

5. Chemical potential: The chemical potential μ is,

$$\mu = \left(\frac{\partial F}{\partial N} \right)_{T,V} = -k_B T \left[\text{In} \frac{V}{N\lambda_T^3} + 1 \right] + Nk_B T \left[\frac{V}{N\lambda_T^3} \cdot \frac{\lambda_T^3}{V} \right]$$
$$= -k_B T \text{ In} \left(\frac{V}{N\lambda_T^3} \right).$$

The previous expressions were much simpler obtained than when calculated in the microcanonical ensemble.

6. Equivalence of ensembles: In the thermodynamic limit the average value of an observable is in general independent of the ensemble (microcanonical or canonical).

$$N \to \infty, V \to \infty, \frac{N}{V} = \text{const.}$$

is taken. One therefore usually chooses the ensemble that is easier to work with.

7. Fluctuations of observables: Fluctuations of observables, $\langle A^2 \rangle - \langle A \rangle^2$, may however be ensemble dependent! An example for an observable for which this is the case is the energy, which is constant, by definition, in the microcanonical ensemble, but distributed according to ($P(E) \sum_{E<E\alpha<E+\Delta} P\alpha \propto \Omega(E) e^{-\beta E}$,) in the canonical ensemble.

Energy Fluctuations

We evaluated the representation ($\frac{C_V}{T} = \left(\frac{\partial S}{\partial T} \right)_v = -\frac{\partial^2 F}{\partial T^2} = \left(\frac{\partial^2}{\partial T^2} \right) (k_B T \text{ In} Z_N)$) for the specific heat in a first step:

$$\frac{C_V}{T} = \frac{\partial^2}{\partial T^2} [k_B T \text{In} Z_n]$$
$$= \frac{\partial}{\partial T} \left[k_B \text{ In } Z_n + \frac{k_B T}{Z_N} \frac{\partial Z_n}{\partial \beta} \frac{\partial \beta}{\partial T} \right] \quad \frac{\partial \beta}{\partial T} = \frac{-1}{k_B T^2}$$

$$= \frac{\partial}{\partial T}\left[k_B \ \text{In}\ Z_n - \frac{1}{TZ_N}\frac{\partial Z_n}{\partial \beta} \right].$$

1. Second derivatives: The remaining derivative with respect to the temperature T are,

$$\frac{\partial}{\partial T} k_B \,\text{In}\, Z_n = \frac{-1}{T^2 Z_N}\frac{\partial Z_n}{\partial \beta}$$

$$\frac{\partial}{\partial T}\frac{-1}{TZ_N}\frac{\partial Z_n}{\partial \beta} = \frac{1}{T^2 Z_N}\frac{\partial Z_n}{\partial \beta} + \left[\frac{1}{TZ_N^2}\left(\frac{\partial Z_n}{\partial \beta}\right)^2 - \frac{1}{TZ_N}\frac{\partial^2 Z_n}{\partial \beta^2} \right]\frac{-1}{k_B T^2}.$$

With the first two terms canceling each other we find,

$$C_V = \frac{1}{k_b T^2}\left[\frac{1}{Z_N}\frac{\partial^2 Z_n}{\partial \beta^2} - \left(\frac{1}{Z_N}\frac{\partial Z_n}{\partial \beta} \right)^2 \right]$$

for the specific heat C_v as a functions of derivatives of the partition function Z_N.

2. Derivatives of the partition function: The definition ($Z_N = \int \frac{d^{3N}q\, d^{3N}}{h^{3N}N!} e^{-\beta H(q,p)}$) for the partition function corresponds to,

$$\frac{1}{Z_N}\frac{\partial Z_n}{\partial \beta} = -\frac{\int \frac{d^{3N}q\, d^{3N}p}{h^{3N}N!}}{\int \frac{d^{3N}q\, d^{3N}p}{h^{3N}N!} e^{-\beta H(q,p)}}, \quad Z_N = \int \frac{d^{3N}q\, d^{3N}p}{h^{3N}N!} e^{-\beta H(q,p)},$$

viz to,

$$\frac{1}{Z_N}\frac{\partial Z_n}{\partial \beta} = -\langle E \rangle, \quad \frac{1}{Z_N}\frac{\partial^2 Z_n}{\partial \beta^2} = \langle E^2 \rangle.$$

Specific heat. Our results of above equation lead to the fundamental relation,

$$C_V = \frac{1}{k_b T^2}\left[\langle E^2 \rangle - \langle E \rangle^2 \right]$$

between the specific heat C_V and the fluctuations $\langle E^2 \rangle - \langle E \rangle^2$ of the energy.

- Both the specific heat $C_V \sim N$ and the right-hand side of the above equation are extensive.
- The specific heat describes the energy exchange between the system and an heat reservoir. It hence makes that sense that CV is proportional to the size of the energy fluctuations.

3. Relative energy fluctuations: The relative energy fluctuations,

$$\frac{\sqrt{\langle E^2 \rangle - \langle E \rangle^2}}{\langle E \rangle} \sim \frac{1}{\sqrt{N}}$$

vanish in the thermodynamic limit $N \to \infty$.

- The scaling relation of above equation if a direct consequence of ($C_V = \frac{1}{k_b T^2}\left[\langle E^2\rangle - \langle E\rangle^2\right]$) and of the fact that both CV and the internal energy U = $\langle E\rangle$ are extensive.
- Equation above is consistent with the demand that the canonical the microcanonical ensembles are equivalent in the thermodynamic limit $N \to \infty$. Energy fluctuations are absent in the microcanonical ensemble.

Paramagnetism

We consider a system with N magnetic atoms per unit volume placed in an external magnetic field $\mathcal{H}$. Each atom has an intrinsic magnetic moment $\mu = 2\mu_0 s$ with spin s = 1/2.

Energy states: In a quantum-mechanical description, the magnetic moments of the atoms can point either parallel or anti-parallel to the magnetic field.

State	Alignment	Moment	Energy	Probability
(+)	parallel to $\mathcal{H}$	$+\mu$	$-\mu\mathcal{H}$	$P_+ = ce^{-\beta\varepsilon} = ce^{+\beta\mu\mathcal{H}}$
(–)	anti-parallel to $\mathcal{H}$	$-\mu$	$+\mu\mathcal{H}$	$P_- = ce^{-\beta\varepsilon} = ce^{-\beta\mu\mathcal{H}}$

We assume here that the atoms interact weakly. One can therefore a single atom as a small system and the rest of the atoms as a reservoir in the terms of a canonical ensemble.

1. Mean magnetic moment: We want to analyze the mean magnetic moment $\langle \mu_{\mathcal{H}}\rangle$ per atom as a function of the temperature T:

$$\langle \mu_{\mathcal{H}}\rangle = \frac{\mu e^{\beta\mu\mathcal{H}} - \mu e^{-\beta\mu\mathcal{H}}}{e^{\beta\mu\mathcal{H}} + e^{-\beta\mu\mathcal{H}}}, \quad \langle \mu_{\mathcal{H}}\rangle = \mu \tanh\frac{\mu_{\mathcal{H}}}{k_b t}$$

where we used that,

$$\tanh y = \frac{e^y - e^{-y}}{e^y - e^{-y}}, \qquad y = \beta\mu\mathcal{H} = \frac{\mu\mathcal{H}}{k_B T}.$$

2. Magnetization: We define the magnetization, i.e. the mean magnetic moment per unit volume, as

$$\langle M\rangle = N\langle \mu_{\mathcal{H}}\rangle$$

and analyze its behavior in the limit of high- and of low temperatures.

High-temperature expansion Large temperatures correspond to $y \ll 1$ and hence to,

$$e^y = 1 + y + \ldots, \quad e^{-y} = 1 - y + \ldots.$$

Then,

$$\tan y = \frac{(1+y+\ldots) - (1-y+\ldots)}{2} \approx y,$$

so that,

$$\langle \mu_{\mathcal{H}} \rangle = \frac{\mu^2 \mathcal{H}}{k_B T}.$$

3. Curie Law: For the magnetic susceptibility χ, defined as $\langle M \rangle = \chi\mathcal{H}$, we then have,

$$\chi = \frac{N\mu^2}{k_B T}.$$

At temperatures high compared to the magnetic energies, $\chi \propto T^{-1}$ which is known as the Curie law.

4. Low-temperature expansion: Low temperatures correspond to $y \gg 1$,

$$e^y \gg e^{-y}, \qquad \tanh y \approx 1,$$

and hence,

$$\langle \mu H \rangle = \mu, \ \langle M \rangle = N\mu.$$

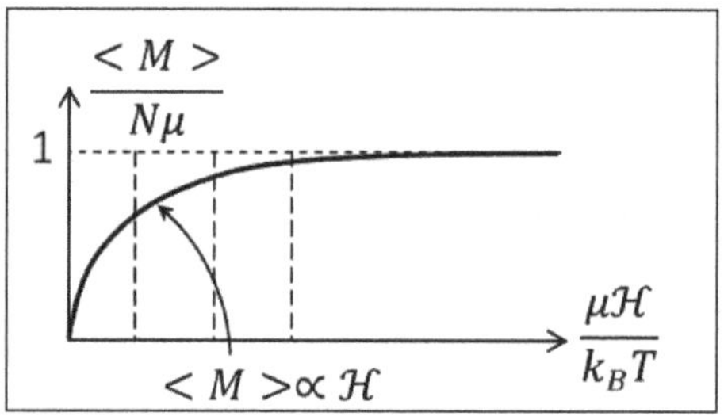

The magnetization saturates at the maximal value at low temperatures independent of $\mathcal{H}$.

Grand Canonical Ensemble

The microcanonical and canonical ensembles and the idea of partition function and laid down a systematic procedure to derive thermodynamic properties and applied theses to study a variety of interesting physical macroscopic systems. In going from microcanonical ensemble to canonical ensemble we relaxed the constraint from complete isolation to allow exchange of energy with the surroundings, making E a variable. Such a system is now a closed system. However, to be more realistic, it is desirable to also allow exchange of matter or more precisely particles constituting the system, making N another variable of the system. System is then said to be an open system. In equilibrium system has an average energy, ⟨E⟩ and an average number of particles ⟨N⟩. This open system is then described by a Grand Canonical Ensemble. This ensemble deals with microstates of a system kept at constant temperature (T), constant chemical potential (μ) in a given volume V. As was seen in the case of canonical ensemble we will now have a new partition function called Grand partition function. This shall be used to define grand potential followed by derivation of thermodynamic properties. We shall apply this approach to study once again the classical ideal gas.

Grand Canonical Distribution Function: System and Heat Bath Approach

We follow the same approach followed for canonical ensemble system and heat bath approach. We take our system (designated A) and put it in a heat bath (designated A′) such that system and the heat bath (designated $A_T = A + A'$), Figure below are allowed to now exchange both energy and particles, keeping system and heat bath in equilibrium at a common temperature T and common chemical potential μ. Heat bath is very large having a large heat capacity. The composite system (A_T) together is completely isolated, such that if at any time if system A has energy E_r and number of particles N_r and heat bath A′ having energy $E_r{}'$ and number of particles $N_r{}'$ then the total energy E_T and total number of particles N_T are constant i.e.

$$E_s + E_s' = E_T = \text{Constant}$$

$$N_r + N_r' = N_T = \text{Constant}$$

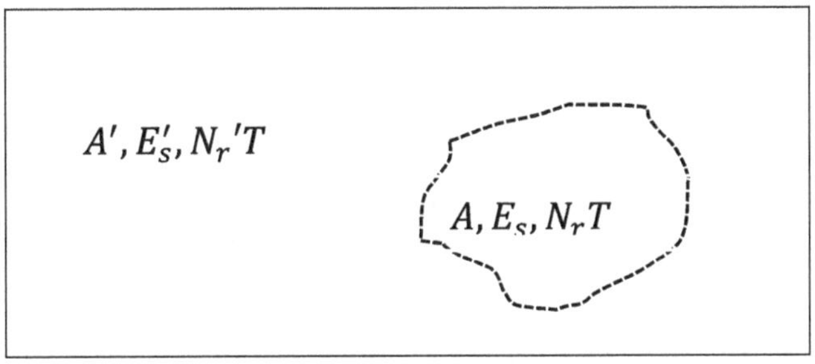

Heat bath A’ at equilibrium with system A at temperature T and chemical potential μ.

Since heat bath is very large, $E_s \ll E_T$ and $N_s \ll N_T$ i.e.

$$\frac{E_s}{E_T} = \left(1 - \frac{E_s'}{E_T}\right) \ll 1$$

$$\frac{N_r}{N_T} = \left(1 - \frac{N_r'}{N_T}\right) \ll 1$$

Since the heat bath is very large the number of states compatible to it with energy E_s' and particles $N_r', \Omega'(E_s', N_r')$, is also very large independent of the nature of the reservoir. It is worth noting that larger the number of states available to the reservoir larger is the probability of it assuming energy E_s' and having particles N_r' and consequently the system immersed in heat bath assuming energy E_s and having number of particles N_r. As per the principle of equal apriori probability, different microstates in the reservoir are equally likely to occur, the probability P_r of system A, occurring with energy E_s and number of particles N_r, is proportional to the number of microstates which the reservoir has i.e. $\Omega'(E_s' N_r')$. Therefore,

$$P_{s,r} \propto \Omega'(E_s', N_r') \equiv \Omega'(E_T - E_s, N_T - N_r)$$

Noting the inequalities ($\frac{E_s}{E_T} = \left(1 - \frac{E_s'}{E_T}\right) \ll 1$) and ($\frac{N_r}{N_T} = \left(1 - \frac{N_r'}{N_T}\right) \ll 1$) we can expand the right hand side around $E'_s = E_T$ and $N'_r = N_T$, which means around $E_s = 0$ and $N_r = 0$ Recalling that Ω' is a very large number which varies very rapidly, it is therefore, more convenient to deal with logarithm of this function which is a slowly varying function. So,

$$\text{In } \Omega'(E'_s N'_r) = \text{In}\,\Omega'(E_T, N_T) + \left(\frac{\partial \text{ In}\,\Omega'}{\partial E'}\right)_{E'_r = E_T}(E'_s - E_T)$$
$$+\left(\frac{\partial \text{ In } \Omega'}{\partial N'}\right)_{E'_s = E_T}(N'_r - N_T) + \ldots$$

Or

$$\text{In } \Omega'(E'_s N'_r) \approx \text{In}\,\Omega'(E_T, N_T) + \left(\frac{\partial \text{ In}\,\Omega'}{\partial E'}\right)_{E'_r = E_T}(-E_s)$$
$$+\left(\frac{\partial \text{ In } \Omega'}{\partial N'}\right)_{N'_r = N_T}$$

Recalling $\left(\frac{\partial \text{ In}\,\Omega'}{\partial E'}\right) = \beta' = \frac{1}{k_B T'}$ and at equilibrium $\beta' = \frac{1}{k_B T'} = \beta = \frac{1}{k_B T}$, $\left(\frac{\partial \text{In}\,\Omega'}{\partial N'}\right) = \alpha' = -\frac{\mu'}{k_B T'} = \alpha = -\frac{\mu}{k_B T}$ equation $\text{In } \Omega'(E'_s N'_r) = \text{In}\,\Omega'(E_T, N_T) + \left(\frac{\partial \text{ In}\,\Omega'}{\partial E'}\right)_{E'_r = E_T}(E'_s - E_T) + \left(\frac{\partial \text{ In } \Omega'}{\partial N'}\right)_{E'_s = E_T}(N'_r - N_T) + \ldots$ can be written as,

$$\text{In } \Omega'(E'_s N'_r) \approx \text{constant} - \beta E_s - \alpha N_r$$

Or

$$\Omega'(E'_s N'_r) \approx \text{constant } e^{\beta E_s - \alpha N_r}$$

From $\left(\frac{N_r}{N_T} = \left(1 - \frac{N'_r}{N_T}\right) \ll 1\right)$ and $\left(\Omega'(E'_s N'_r) \approx \text{constant } e^{\beta E_s - \alpha N_r}\right)$ we get,

$$P_{s,r} \propto e^{-\beta(E_s - \mu N_r)}$$

Since sum of all probabilities must sum to 1, we get,

$$P_{s,r} = \frac{e^{-\beta(E_s - \mu N_r)}}{\sum_{s,r} e^{-\beta(E_s - \mu N_r)}}$$

Where, summation is over all the (E_s, N_r) states which system A can take. Equation above is the Grand canonical distribution function which provides the probability of the system in state with energy E_s and number of particles N_r.

Furthermore, since equation $P_{s,r} = \frac{e^{-\beta(E_s-\mu N_r)}}{\sum_{s,r} e^{-\beta(E_s-\mu N_r)}}$ has been derived by using very fundamental ideas of physics involving conservation of energy, extensiveness of energy $E_T = E_r + E'_r$, conservation of number of particles, extensiveness of number of particles and slow and smooth variation of $\ln \Omega(E,N)$, independent of the choice of quantum mechanics or classical mechanics.

Grand Canonical Distribution Function: Most Probable Value Approach

We consider an ensemble of $\mathbb{N}$ identical systems which are allowed to mutually share total energy $\mathbb{N} < E >$ and a total number of particles $\mathbb{N} < N >$. Let n_{sr} denote the number of system which have at any instant t an amount of energy E_s and number of particles N_r then these together satisfy the following three conditions:

$$\sum_{s,r} n_{sr} = \mathbb{N}$$

$$\sum_{r} n_{sr} E_s = \mathbb{N} < E >$$

$$\sum_{r} n_{sr} N_r = \mathbb{N} < N >$$

Any set of values $\{n_{sr}\}$which satisfy conditions stated in equations above is a possible way of distributing energy and particles among the members of the ensemble. Number of such possible ways $\{n_{rs}\}$, of distributing energy and particles among the members of the ensemble is given by,

$$\mathbb{W}\{n_{sr}\} = \frac{\mathbb{N}!}{\prod_{n_{sr}} (n_{sr}!)}$$

According to principle of equal apriori probability, all possible distribution of energy and particles among the members of the ensemble are equally likely, the frequency of achieving a given set $\{n_{rs}\}$ is directly proportional to the number $\{n_{rs}\}$. Out of all these sets the most probable set shall be the one for which $\{n_{rs}\}$ is maximum. Once we have identified this set say $\{n^*_{rs}\}$, we shall be physically interested in this distribution. However statistically we can not ignore other $\{n_{rs}\}$'s with their corresponding weights or frequencies to calculate the expectation values or mean values of the numbers n_r as given below:

$$\langle n_{rs} \rangle = \frac{\sum'_{\{n_{sr}\}} n_r \mathbb{W}\{n_{sr}\}}{\sum'_{\{n_{sr}\}} \mathbb{W}\{n_{sr}\}}$$

Significance of n^*_{rs} lies in the fact that the fraction $\frac{n^*_{sr}}{\mathbb{N}}$ is the grand canonical probability P_{sr}.

Let us now calculate the desirable numbers $\{n^*_{rs}\}$ and $\langle n_{sr} \rangle$. Again for the reasons stated earlier In $\mathbb{W}$ is a slowly varying function and we shall use it to proceed further, which implies,

$$\ln \mathbb{W} = \ln \mathbb{N}! - \sum_{sr} \ln n_{sr}!$$

In thermodynamic limit apply Stirlings formula and equation $\sum_{s,r} n_{sr} E_s = \mathbb{N} < E >$ becomes,

$$\text{In } \mathbb{W} = \mathbb{N} \text{ In } \mathbb{N} - N - \sum_{sr} n_{sr} \text{ In} n_{sr} + \sum_{sr} n_{sr}$$

$$= \mathbb{N} \text{ In } \mathbb{N} - \sum_{r} n_{sr} \text{ In } n_{sr}$$

If n_{sr} changes by a small amount In $\mathbb{W}$ also changes by a small amount as given below:

$$\delta(\text{In } \mathbb{W}) = \sum_{r} (\text{In } n_{sr} + 1)\delta n_{sr}.$$

Now for n_{sr} to be maximum $\delta(\text{In } \mathbb{W})$ should become zero, provided equations $\sum_{s,r} n_{sr} = \mathbb{N}$ and $\sum_{r} n_{sr} E_s = \mathbb{N} < E >$ and $\sum_{r} n_{sr} N_r = \mathbb{N} < N >$ also satisfy the following conditions simultaneously,

$$\sum_{r} \delta n_{sr} = 0$$

$$\sum_{sr} E_s \ \delta n_{sr} = 0$$

$$\sum N_r \ \delta n_{sr} = 0$$

Now to get the set $\{n_r^*\}$, method of Lagrange's undetermined multiplier can be used according to which equations above together yield the condition,

$$\sum_{r} \left(-(\text{In} n_{sr} + 1) - \alpha N_r - \beta E_s - \gamma\right)\delta n_{sr} = 0$$

Where α, β and γ are Lagrange's undetermined multipliers. Since δn_r is completely arbitrary, for equation $\text{In } \mathbb{W} = \mathbb{N} \text{ In } \mathbb{N} - N - \sum_{sr} n_{sr} \text{ In} n_{sr} + \sum_{sr} n_{sr} = \mathbb{N} \text{ In } \mathbb{N} - \sum_{r} n_{sr} \text{ In } n_{sr}$ to be satisfied its coefficients must become zero, i.e. for all r,

$$\left(-(\text{In } n_r^* + 1) - \alpha N_r - \beta E_s - \gamma\right) = 0$$

Or,

$$\text{In} n_{sr}^* = -1 - \alpha N_r - \beta E_s - \gamma$$

Or,

$$n_{sr}^* = e^{-(1+\gamma)} e^{-\beta Er - \alpha Nr} = C e^{-\beta E_r - \alpha N_r}$$

Where now C, α and β are two undetermined parameters and above equation gives us the most probable distribution. To calculate C, we note,

$$\sum_{r} n_{sr}^* = N = C \sum_{sr} e^{-\beta E_s - \alpha N_r}$$

or

$$C = \frac{\mathbb{N}}{\sum_{sr} e^{-\beta E_s - \alpha N_r}}.$$

Therefore,

$$P_{sr} = \frac{n_{sr}^*}{\mathbb{N}} = \frac{e^{-\beta E_s - \alpha N_r}}{\sum_{sr} e^{-\beta E_s - \alpha N_r}}$$

In Thermodynamic limit at equilibrium $\frac{n_{sr}^*}{\mathbb{N}} \approx \frac{\langle n_{sr} \rangle}{\mathbb{N}}$, therefore,

$$\frac{\langle n_{sr} \rangle}{\mathbb{N}} \approx \frac{n_{sr}^*}{\mathbb{N}} = \frac{e^{-\beta E_s - \alpha N_r}}{\sum_{sr} e^{-\beta E_s - \alpha N_r}}$$

The undetermined parameters α and β are determined by the equations for average number of particles ⟨N⟩ and average energy ⟨E⟩ in each system of the ensemble with prior knowledge of their values:

$$\langle N \rangle = \frac{\sum_{sr} N_r e^{-\beta E_s - \alpha N_r}}{\sum_{sr} e^{-\beta E_s - \alpha N_r}} = -\frac{\partial}{\partial \alpha}\left\{ \text{In} \sum_{sr} N_r e^{-\beta E_s - \alpha N_r} \right\}$$

$$\langle E \rangle = \frac{\sum_{sr} E_s e^{-\beta E_s - \alpha N_r}}{\sum_{sr} e^{-\beta E_s - \alpha N_r}} == -\frac{\partial}{\partial \beta}\left\{ \text{In} \sum_{sr} E_r e^{-\beta E_s - \alpha N_r} \right\}$$

Link between Grand Canonical Ensemble and Thermodynamics: Thermodynamic Potential

To establish a link between the Grand canonical ensemble and thermodynamics, we introduce a quantity q, called q potential as follows:

$$q = \text{In}\left\{ \sum_{sr} e^{-\beta E_s - \alpha N_r} \right\}$$

q is a function of α, β and all the energy states E_s. Taking the differential of q, we have,

$$dq = -\frac{\sum_{sr} N_r e^{-\beta E_s - \alpha N_r}}{\sum_{sr} e^{-\beta E_s - \alpha N_r}} d\alpha - \frac{\sum_{sr} E_s e^{-\beta E_s - \alpha N_r}}{\sum_{sr} e^{-\beta E_s - \alpha N_r}} d\beta$$

$$-\beta \sum_{sr} \frac{e - \beta E_s - \alpha N_r}{\sum_{sr} e - \beta E_s - \alpha N_r} dE_s$$

Or using equations $\frac{\langle n_{sr} \rangle}{\mathbb{N}} \approx \frac{n_{sr}^*}{\mathbb{N}} = \frac{e^{-\beta E_s - \alpha N_r}}{\sum_{sr} e^{-\beta E_s - \alpha N_r}}$, $\langle N \rangle = \frac{\sum_{sr} N_r e^{-\beta E_s - \alpha N_r}}{\sum_{sr} e^{-\beta E_s - \alpha N_r}} = -\frac{\partial}{\partial \alpha}\left\{ \text{In} \sum_{sr} N_r e^{-\beta E_s - \alpha N_r} \right\}$ and $\langle E \rangle = \frac{\sum_{sr} E_s e^{-\beta E_s - \alpha N_r}}{\sum_{sr} e^{-\beta E_s - \alpha N_r}} = -\frac{\partial}{\partial \beta}\left\{ \text{In} \sum_{sr} E_r e^{-\beta E_s - \alpha N_r} \right\}$, we have

$$dq = -\langle N \rangle d\alpha - \langle E \rangle d\beta - \frac{\beta}{\mathbb{N}} \sum_{sr} \langle n_{sr} \rangle dE_s.$$

Which can be further written as,

$$d\left(q+\alpha\langle N\rangle+\beta\langle E\rangle\right)=\beta\left(-\frac{\alpha}{\beta}d\langle N\rangle-d\langle E\rangle-\frac{1}{\mathbb{N}}\sum_{sr}\langle n_{sr}\rangle dE_s\right)$$

Now if we compare the expression with in the parenthesis with the first law of thermodynamics:

$$\delta Q=-\frac{\alpha}{\beta}d\langle N\rangle-d\langle E\rangle+\delta W$$

We can arrive at the following correspondence,

$$\mu=-\frac{\alpha}{\beta}$$

$$\delta W=-\frac{1}{\mathbb{N}}\sum_{sr}\langle n_{sr}\rangle dE_s$$

So that,

$$d\left(q+\alpha\langle N\rangle+\beta\langle E\rangle\right)=\beta\delta Q$$

A closer look at equation above and comparing it with $dS=\frac{\delta Q}{T}$ tells us that the coefficient β on the right hand side should correspond to reciprocal of temperature T, so that,

$$\beta=\frac{1}{k_B T}$$

From equation $\mu=-\frac{\alpha}{\beta}$, it immediately follows that,

$$\alpha=-\frac{\mu}{k_B T}$$

Left hand side of equation, then provides correspondence of the quantity $q+\alpha\langle N\rangle+\beta\langle E\rangle$ with entropy, such that,

$$\frac{S}{k_B}=q+\alpha\langle N\rangle+\beta\langle E\rangle$$

Or

$$q=\frac{S}{k_B}-\alpha\langle N\rangle-\beta\langle E\rangle=\frac{S}{k_B}-\frac{\mu\langle N\rangle}{k_B T}-\frac{\langle E\rangle}{k_B T}=\frac{TS+\mu\langle N\rangle-\langle E\rangle}{k_B T}$$

The function $k_B T q=TS+\mu\langle N\rangle-\langle E\rangle$ is the well known Thermodynamic Potential Φ which describes an open, isochoric and isothermal collection,

$$\Phi(T,\mu,V)=TS+\mu\langle N\rangle-\langle E\rangle=q k_B T$$

Recalling that Gibbs free energy $G=\mu\langle N\rangle=\langle E\rangle-TS+PV$,

equation $q=\frac{S}{k_B}-\alpha\langle N\rangle-\beta\langle E\rangle=\frac{S}{k_B}-\frac{\mu\langle N\rangle}{k_BT}-\frac{\langle E\rangle}{k_BT}=\frac{TS+\mu\langle N\rangle-\langle E\rangle}{k_BT}$ becomes,

$$q=\text{In}\left\{\sum_{sr}e^{-\beta E_s-\alpha N_r}\right\}=\frac{PV}{k_BT}$$

Equation above provides a bridge between thermodynamics and grand canonical ensemble.

Grand Canonical Partition Function and Thermodynamic Functions

We can define grand canonical partition function Z,

$$\mathbb{Z}=\left\{\sum_{sr}e^{-\frac{E_s}{k_BT}+\frac{\mu N_r}{k_BT}}\right\}$$

Defining $e\frac{\mu}{k_BT}=f$ called fugacity equation above can be written as,

$$\mathbb{Z}=\left\{\sum_{sr}(f)^{N_r}e^{-\frac{E_s}{k_BT}}\right\}=\left\{\sum_{r}(f)N_r\left(\sum_{s}e^{-\frac{E_s}{k_BT}}\right)\right\}$$

The expression $ZN_r(V,T)=\left(\sum_s e^{-\frac{E_s}{k_BT}}\right)$ is the canonical partition function for a fixed N_r and has in it via energy, dependence on volume. Therefore, Grand canonical partition function can be written as,

$$\mathbb{Z}=\left\{\sum_{N_r=o}^{\infty}(f)^{N_r}Z_{N_r}(V,T)\right\}$$

Hence q is related to the partition function via the relation,

$$q=\text{In }\mathbb{Z}(f,V,T)=\frac{PV}{k_BT}.$$

Now we are in a position to write down the thermodynamic functions of interest in terms of q or Z.

1. Pressure:

$$P=\frac{k\ T}{}q(f,V,T)=\text{In}\quad(f,V,T)$$

Average number of particles in equilibrium,

$$N=\langle N\rangle=k_BT\left[\frac{\partial q(\mu,V,T)}{\partial\mu}\right]_{V,T}=\left[f\frac{\partial q(f,V,T)}{\partial f}\right]_{V,T}$$

2. Internal Energy:

$$E = \langle E \rangle = -\left[\frac{\partial q(f,V,T)}{\partial \beta}\right]_{V,f} = k_B T^2 \left[\frac{\partial q(f,V,T)}{\partial T}\right]_{V,f}.$$

One can get E as a function of N, V, T by eliminating f between equations

$$N = \langle N \rangle = k_B T\left[\frac{\partial q(\mu,V,T)}{\partial \mu}\right]_{V,T} = \left[f\frac{\partial q(f,V,T)}{\partial f}\right]_{V,T} \text{ and } E = \langle E \rangle = -\left[\frac{\partial q(f,V,T)}{\partial \beta}\right]_{V,f} = k_B T^2 \left[\frac{\partial q(f,V,T)}{\partial T}\right]_{V,f}.$$

Using this functional form, taking derivative with respect to T gives specific heat at constant volume.

3. Equation of state: It is a relationship between P, V, T and can be obtained by eliminating f between equations $P = \frac{k_B T}{V} q(f,V,T) = \text{In } \mathbb{Z}(f,V,T)$ and $N = \langle N \rangle = k_B T\left[\frac{\partial q(\mu,V,T)}{\partial \mu}\right]_{V,T} = \left[f\frac{\partial q(f,V,T)}{\partial f}\right]_{V,T}$.

4. Helmholtz free energy:

$$F = E - TS + PV - PV = G - PV = N\mu - PV$$

$$= Nk_B T \text{ In } f - k_B T \text{ In } \mathbb{Z}(f,V,T) = -k_B T \text{ In} \frac{\mathbb{Z}(f,V,T)}{f^N}.$$

Interestingly, argument of the logarithm is canonical distribution function.

5. Entropy:

$$S = \frac{E-F}{T} = k_B T\left[\frac{\partial q(f,V,T)}{\partial T}\right] - Nk_B \text{ In} f + k_B q(f,V,T).$$

Grand Canonical Partition Function and Single Orbital Partition Functions

Keeping equation $\sum_r n_{sr} E_s = N < E >$ in mind the grand partition function $\mathbb{Z} = \left\{\sum_{N_r=0}^{\infty} (f)^{N_r} Z_{N_r}(V,T)\right\}$,

$$\mathbb{Z} = \left\{\sum_{N_r=0}^{\infty} (f)^{N_r} Z_{N_r}(V,T)\right\}$$

By recalling that,

$$Z_{N_r}(V,T) = \left(\sum_s e - \frac{E_s}{k_B T}\right) = \left(\sum_{\{N_r\}} e^{-\sum_r \frac{N_r E_r}{k_B T}}\right)$$

With $\sum_{\{N_r\}} N_r = N$, can be written as,

$$\mathbb{Z} = \left\{\sum_{N_r=0}^{\infty} (f)^{N_r} \sum_{\{N_r\}} \left(e^{-\sum_r \frac{N_r E_r}{k_B T}}\right)\right\}$$

Or,

$$\mathbb{Z} = \left\{ \sum_{N_r=o}^{\infty} \sum_{\{N_r\}} (f)^{\sum_r N_r} \left(e^{-\sum_r \frac{N_r E_r}{k_B T}} \right) \right\}$$

Or,

$$\mathbb{Z} = \left\{ \sum_{N_r=o}^{\infty} \sum_{\{N_r\}} \prod_r \left(fe - \frac{E_r}{k_B T} \right)^{N_r} \right\}$$

This can be further written as,

$$\mathbb{Z} = \left\{ \prod_r \sum_{N_r=o}^{\infty} \left(fe^{-\frac{E_r}{k_B T}} \right)^{Nr} \right\}$$

Here the product is over the partition function for a single orbital only with all possible number of particles. In this form grand particle function becomes a very useful result providing us to find the grand partition function from one orbital partition function.

Liouville's Theorem

There are two approaches to thermodynamics:

- Phenomenological (or macroscopic).
- Fundamental (or microscopic).

In both approaches we make assumptions (low energy density, equilibrium, etc.) and hope that the nature will respect them at least in some limit. An active area of research is to explore other limits (high energy density, equilibrium) under other assumptions. Now that we have seen a phenomenological formulation of thermodynamics we will switch to a derivation of thermodynamics from a more fundamental viewpoint.

Consider again a large system with 3N degrees of freedom (d.o.f.). For example, 6.02214×10^{23} molecules in a box. Microstate of the system can be described by a 6N dimensional vector (q_1, q_2,..., q_{3N}, p_1, p_2,..., p_{3N}) in phase space, where {q_1, q_2,..., q_{3N} } are the position coordinates and {p_1, p_2,..., p_{3N}} are the momentum coordinates. Trajectories in phase space described by Hamiltonian equations of motion:

$$\dot{p}_i = -\frac{\sigma H}{\sigma q_i}$$

and,

$$\dot{q}_i = \frac{\sigma H}{\sigma p_i},$$

where i = 1...3N. Consider representative points in an infinitesimal 6N dimensional hypercube with volume:

$$d\Gamma = dq_1 ... dq_{3N}\ dp_1 ... dp_{3N}$$

In time δt i the representative point move to new locations,

$$q_i' = qi + \dot{q}_i dt$$

$$p_i' = pi + \dot{p}_i dt$$

and the volume becomes distorted,

$$d\Gamma' = \left(dq_1 + \frac{\partial \dot{q}_i}{\partial q_1} dq_1 dt\right) ... \left(dq_{3N} + \frac{\partial \dot{q}_{3N}}{\partial q_{3N}} dq_{3N} dt\right)\left(dp_1 + \frac{\partial \dot{p}_1}{\partial p_1} dp_1 dt\right) ... \left(dp_{3N} + \frac{\partial \dot{p}_{3N}}{\partial p_{3N}} dp_{3N} dt\right).$$

By expanding to the linear order in dt we obtain,

$$\frac{d\Gamma' - d\Gamma}{dt d\Gamma} = \sum_{i=1}^{3N}\left(\frac{\partial \dot{q}_i}{\partial q_1} + \frac{\partial \dot{p}_1}{\partial p_1}\right) = \sum_{i=1}^{3N}\left(\frac{\partial}{\partial q_i}\frac{\partial H}{\partial p_i} - \frac{\partial}{\partial p_i}\frac{\partial H}{\partial q_i}\right) = 0$$

By Hamiltonian equations of motion. Alternatively, the divergence of the velocity flow v = $(\dot{q}1, \ldots, \dot{q}_{3N}, \dot{p}_1, \ldots, \dot{p}_{3N})$ vanishes, i.e.

$$\nabla \cdot v\left(\frac{\partial}{\partial q_1}, \ldots, \frac{\partial}{\partial q_{3N}}, \frac{\partial}{\partial p_1}, \ldots, \frac{\partial}{\partial p_{3N}}\right) \cdot \left(\dot{q}_1, \ldots, \dot{q}_{3N}, \dot{p}_1, \ldots, \dot{p}_{3N}\right) = \sum_{i=1}^{3N}\left(\frac{\partial}{\partial q_i}\frac{\partial H}{\partial p_i} - \frac{\partial}{\partial p_i}\frac{\partial H}{\partial q_i}\right) = 0$$

This is the famous Liouville's theorem which says that the flow in a phase space of Hamiltonian systems is incompressible, i.e. $d\Gamma = d\Gamma'$.

The Liouville's theorem can also be expressed in a dierential form. Let us define a multiparticle distribution function $\rho(q_1, \ldots, q_{3N}, p_1, \ldots, p_{3N})$ over the phase space, where

$$\rho(q_1, \ldots, q_{3N}, p_1, \ldots, p_{3N}) dq_1 \ldots dq_{3N}\ dp_1 \ldots dp_{3N}$$

is a probability to find a system in a state between $(q_1, \ldots, q_{3N}, p_1, \ldots, p_{3N})$ and $(q_1 + dq_1, \ldots, q_{3N}\ dq_{3N}, p_1 + dp_1, \ldots, p_{3N})$. Then the continuity equation,

$$\frac{dp}{dt} = \frac{\partial \rho}{\partial t} + \frac{\partial}{\partial q_i}(\dot{q}i\rho) + \frac{\partial}{\partial p_i}\left(\dot{p}i\rho\right) = 0$$

where the Poisson brackets are defined as,

$$\{A, B\} = \sum_{i=1}^{3N}\left(\frac{\partial A}{\partial q_i}\frac{\partial B}{\partial p_i} - \frac{\partial A}{\partial p_i}\frac{\partial B}{\partial q_i}\right).$$

The averages are defined by,

$$\langle \mathcal{O} \rangle = \int \mathcal{O}(q, p)\rho(q, p\,t) dq dq$$

and their evolution,

$$\frac{d\langle \mathcal{O}\rangle}{dt}\int \mathcal{O}(q,p)\frac{\partial\rho(q,pt)}{\partial t}dqdq = -\int \mathcal{O}\sum_{i=1}^{3N}\left(\frac{\partial\rho}{\partial q_i}\frac{\partial H}{\partial p_i}-\frac{\partial\rho}{\partial p_i}\frac{\partial H}{\partial q_i}\right)dqdp =$$

$$=\int \rho\sum_{i=1}^{3N}\left(\frac{\partial\rho}{\partial q_i}\left(\mathcal{O}\frac{\partial H}{\partial p_i}\right)-\frac{\partial\rho}{\partial p_i}\left(\mathcal{O}\frac{\partial H}{\partial q_i}\right)\right)dqdp =$$

$$=\int \rho(q,p)\sum_{i=1}^{3N}\left(\frac{\partial\mathcal{O}}{\partial q_i}\frac{\partial H}{\partial p_i}-\frac{\partial\mathcal{O}}{\partial p_i}\mathcal{O}\frac{\partial H}{\partial q_i}-\mathcal{O}\left(\frac{\partial}{\partial q_i}\frac{\partial H}{\partial p_i}-\frac{\partial}{\partial p_i}\mathcal{O}\frac{\partial H}{\partial q_i}\right)\right)dqdp = \langle\{\mathcal{O},H\}\rangle$$

Thus, for the ensemble average not to change with time (or to be in equilibrium) we must have,

$$\frac{\partial eq}{\partial t} = -\{\rho_{eq}, H\} = 0.$$

For the distributions which are not in equilibrium very little is known. First of all there are two types of averages: ensemble average ($\langle\mathcal{O}\rangle = \int \mathcal{O}(q,p)\rho(q,pt)dqdq$), and time average,

$$\langle\mathcal{O}\rangle_t = \frac{1}{t}\int \mathcal{O}(q(t),p(t))dt$$

Ergodicity is the assumption that the two averages are equivalent which was proved for only relatively simple systems, but is believed to hold for more complicated systems as well. Of course, the exact specification of the state might not be possible in experiments with a large number of d.o.f. It is convenient instead to define a single particle correlation function,

$$f(q,\ p) = N\int \rho(q_1 = q,\ p_1 = p,\ q_2,\ p_2,\ \ldots,\ q_N,\ p_N)dq_2dp_2\ldots dq_Ndp_N$$

The main idea in the kinetic theory is to make some assumptions about unobservable microscopic information (e.g. q's and p's), and to concentrate on only observable macroscopic information (e.g. V and P). Thus, we make two simplifying assumptions:

1. Molecular chaos assumption:

$$\rho(q_1, \ldots, q3_N,\ p_1,\ \ldots,\ p3_N) = f(q_1,\ p_1)\ f(q_2,\ p_2)\ldots f(q_N,\ p_N)$$

where f(q, p) is a single particle distribution function, where,

$$f(q,p)dqdp$$

is a probability to find a particle in a state between (q,p) and $(q+dq,\ p+dp)$.

2. Separation of scales assumption:

Size of Box ≫ Resolution Scale ≫ Inter-molecular Separation ≫ Molecular Size The main problem in a kinetic theory is to find f(q, p, t) for a given type of interaction and external force F.

Ergodic Hypothesis

The microcanonical ensemble $d\xi$ is ergodic with respect to the Hamiltonian dynamics. if the ergodic hypothesis holds, then we have a precise mechanical definition of macroscopic equilibrium $d\xi$, the existence a.e. of time averages of integrable functions f representing macroscopic observables, the equality of such averages with space averages and a justification of the adoption of the microcanonical ensemble. In summary, the statistical mechanics prescription would be justified.

Success and Nonergodicity

Before going on, we present an example of a Hamiltonian system that is not ergodic. Consider two independent harmonic oscillators; since the phase space can be decomposed in two independent parts, one for each oscillator and both of nonzero $d\xi$ measure, the pair (T^t, ξ) is not ergodic. The argument easily generalizes to noninteracting systems with finitely many particles, so that for the ergodicity of (T^t, ξ) the interaction among its particles is fundamental.

For the so-called gas of hard spheres whose particles interact only via elastic collisions, it was recently demonstrated by Sim´anyi that elastically colliding $N \geq 2$ hard balls, of the same radius and arbitrary masses, on the flat torus of any finite dimension is ergodic (also mixing, see ahead); the energy is fixed and the total momentum is zero. This is a generalization of results by Sinai around 1970 for $N = 2$. There are some classes of two-particle mechanical systems, i.e., the Hamiltonian is the sum of the kinetic energy K and the potential energy U, on the two-dimensional torus studied by Donnay and Liverani, whose potential U is radially symmetric and vanishes outside a disk, for which ergodicity is also proved. These can be considered the most realistic models in statistical mechanics where the ergodic hypothesis has been rigorously established. What is then the attitude when other interactions are present? Unfortunately a satisfactory answer is still missing. Some researchers (as in most textbooks) suggest that the postulate of equal a priori probability should be invoked and taking the experimental results as a final confirmation.

Some studies in the area have argued that ergodic theory did not success in explaining the positive results of statistical mechanics. For instance, Earman and R´edei claim that many models for which statistical mechanics works are likely not ergodic and so nonergodic properties must be invoked. As an alternative they have proposed that an ergodic-like behavior, i.e., the validity of Eq. $f^*(\xi) = \langle f \rangle$ should hold only for a finite set of observables f, where it is proposed to restrict ergodicity and mixing to suitable variables). They assert that for each model the equilibrium statistical mechanics predicts values for only a finite set of observables, and one should investigate whether the ergodic-like behavior holds only for such set, which may differ from system to system.

There are some criticisms on the assumption that the outcome of a measurement can be described by infinite time averages; for example, only results concerning quantities in equilibrium could be obtained in this way. Note that in this work only equilibrium statistical mechanics is considered. Also that sets of Lebesgue measure zero can be neglected has the opposition of some researchers;

Another suggestive argument employed for the justification of the microcanonical ensemble comes from analogies with the second law of thermodynamics. Separate the phase space Γ into a finite number of M disjoint cells and associate a probability pj of a representative of the microstate be in the j-th cell; then we have the constrain $\sum_{j=1}^{M} p_j = 1$. Such separation of phase space into cells of nonzero volume is usually called a coarse grain partition and associated to each of them one defines the Gibbs coarse grain entropy,

$$S(p_j) = -\sum_{j=1}^{M} p_j \ln p_j.$$

Now impose that the equilibrium is attained for the distribution of p_j with maximum entropy under the above constrain; it is left as an exercise to conclude that such maximum is obtained for $p_j = 1/M$, for all j, that is, equal probability. In the formal limit of infinite many cells (with vanishing sizes) one gets an indication for the validity of the equal a priori probability. Note, however, that the microscopic evolution does not enter explicitly in the argument, as should be expected, and that this argument only shift the actual problem to another one.

Since the idea of separating the phase space into finitely many cells was mentioned, it is worth recalling Boltzmann reasoning that led him to the "original formulation of the ergodic hypothesis" in the 1870's. Under time evolution Boltzmann supposed that the cells are cyclically permuted; then it became natural to assume that time averages could be performed by averaging over cells. In the limit of infinite cells one would get Eq $f^*(\xi) = \langle f \rangle$.

There is a clear general uncomfortable reaction in the literature with respect to the missing proofs of the ergodic hypothesis for a large class of particle interactions. In some opinion such great difficulty for ergodic proofs should be expected due to the possibility of coexistence of different phases. More precisely, suppose that (T^t, $d\xi$) is not ergodic; another mathematical result says that every invariant measure is a convex combination of ergodic ones. In case $d\xi = \lambda d\mu_1 + (1 - \lambda)d\mu_2$, with $0 < \lambda < 1$ and (T^t, μ_1) and (T^t, μ_2) ergodic with $\mu_1 \perp \mu_2$, then, given an observable f, there are two disjoint sets $A_1, A_2 \subset \Gamma$ with $\mu_1(A_1) = 1 = \mu_2(A_2)$ and $\mu_1(A_2) = 0 = \mu_2(A_1)$, so that by Birkhoff theorem and ergodicity, time averages of f exist for initial conditions $\xi \in A_j$ and resulting in $R\Gamma\, f(\xi)\, d\mu_j(\xi)$, j = 1. That is, there are two different important equilibrium for the system, one for each ergodic component μ_1, μ_2, which one can interpret as the coexistence of different phases. The argument generalizes for more than two ergodic measures in the decomposition. From this point of view, the results of Sinai and Sim´anyi can be interpreted as a proof of just one phase for the gas of finitely many hard spheres. This is closely related to a rigorous approach to the description of phase transitions when one works directly with infinite volume systems, in which the corresponding thermodynamic limit is taken on probability measures.

Without going into a detailed discussion of nonequilibrium, we observe that the usual justification why systems approach to equilibrium under time evolution is via the property of mixing, which was intuitively explored by Gibbs (through the well-known example of stirring a drop of ink in a liquid), but its mathematical definition is due to von Neumann in 1932: (τ^t, μ) is mixing (μ invariant) if for any $A, B \subset \Omega$ one has,

$$\lim_{t\to\infty} \mu(\tau^t(A) \cap B) = \mu(A)\mu(B)$$

This expression means that as time increases the portion of a given measurable set A that then resides any set B is proportional to the measure of B; thus, according to the probability measure

μ, τ^t (A) becomes uniformly distributed over Ω. If this condition is satisfied then, necessarily, the system is ergodic; in fact, if $\tau^t(A) = A$ and by taking B = A in ($\lim_{t\to\infty}\mu(\tau^t(A)\cap B) = \mu(A)\mu(B)$), it follows that $\mu(A) = \mu(A)^2$, so that $\mu(A) = 0$ or $\mu(A) = 1$, i.e., (τ^t, μ) is ergodic. Therefore, the natural condition claimed to assure convergence to equilibrium implies ergodicity. Although this proof sounds simple today, it hides a complex set of developments and such implication was not known by Boltzmann and Gibbs, and was first proved many years after their works on the foundations of statistical mechanics. Certainly, it is usually harder to show that a system is mixing than its ergodicity (in case such properties hold).

Lombardi argued that although some researchers take a pragmatic position to the microcanonical ensemble, in accepting that mixing of $d\xi$ plays a significant role in the description of the approach to macroscopic equilibrium $d\xi$ then, according to the preceding paragraph, ergodicity of $d\xi$ comes up again. Ne-vertheless, the necessity of mixing for evolution toward equilibrium and irreversibility is not consensus.

A system that is ergodic but not mixing is the rotation of the circle S. Let τ^t: S $\rightarrow$ S be defined by $\tau^t(x) = x + \alpha t$, mod 2π, $x \in [0, 2\pi]$, for some irrational number $0 < \alpha < 1$ (recall that x mod 2π means that for any integer number n the numbers x and $x + n2\pi$ are identified and one picks up the representative in the interval $[0, 2\pi)$). One can show that τ^t with the Lebesgue measure dx is ergodic, however it is not mixing; in fact, for $0 < a \ll 1$ set $A = [0, a]$ and $B = [\pi, \pi + a]$, which are two disjoint intervals in S, then $\tau^t(A) = [\alpha t, a+\alpha t]$ is a rotation, and occurrences of empty and nonempty intersections with B alternate sequentially, so that $\mu(T^t(A) \cap B)$ has no limit as $t \rightarrow \infty$. In other words, the rotation does not spread sets (in order to be mixing) although it "spreads sets in average" (the essential difference between ergodicity and mixing is the presence of the time average in the former).

Completely Integrable × Ergodic Systems

Another source of indications of the prevalence of nonergodic Hamiltonian systems (with respect to $d\xi$) are the perturbations of the so-called completely integrable Hamiltonian systems. Such systems are those that have as many independent constants of motion (like energy) as degrees of freedom; a combination of results by Liouville and Arnold shows that the solutions of Hamiltonian equations can be explicitly found (up to evaluation of certain integrals and inverse functions) and in case the accessible phase space is compact it can be foliated by invariant tori (whose dimension coincides with the number of degrees of freedom) and each orbit are just rotations in a torus. Hence, in this case the motion is described in a simple way. Think of the harmonic oscillator with one degree of freedom whose phase space can be foliated by its orbits, and these orbits are (continuously deformed) one dimensional tori. Most textbook examples of mechanical systems are completely integrable.

Clearly complete integrable systems with more than one degree of freedom do not satisfy the ergodic hypothesis. For a long time it was conceived that a small nonlinear perturbation of an integrable system would make them ergodic, so that aside from the exceptions of completely integrable cases, Hamiltonian systems were ergodic (at least for "most" compact energy surfaces). However, in one of the first computer simulations in physics performed by Fermi, Pasta and Ulam on anharmonic perturbations of a chain of independent harmonic oscillators (so an integrable system) in 1955, indications were found that the resulting systems were not ergodic.

A rigorous approach for the lack of ergodicity of perturbations of completely integrable systems is one of the main contents of the KAM theorem. This theorem has shown that weak perturbations of integrable systems just slightly distort most of tori ("most" means of large Lebesgue measure), which are persistent and the resulting systems are not ergodic. For a more precise formulation and Refs. Although the KAM theorem has nothing to say in the complement set of those distorted tori, a huge amount of numerical simulations have revealed "chaotic behavior" and dense trajectories in this remaining region of phase space, and even ergodic behavior for some sufficiently large perturbations.

Researchers have presented an interesting result on generic Hamiltonian systems; generic means that the set in question is constructed as a countable intersection of open dense sets (and so it is also dense by Baire theorem). For example, in the interval the set of irrational numbers is generic, while the set of rational numbers is not. Besides historical remarks on integrability and ergodicity in Hamiltonian systems, they have shown that generic (infinitely) differentiable Hamiltonian systems are neither ergodic nor completely integrable. The proof is actually related to the KAM theorem.

What about most of the systems of interest to statistical physics? And if one restricts the consideration to differentiable Hamiltonians H = K + U, that is, to mechanical systems whose Hamiltonian is the sum of the kinetic energy K and the potential energy U (in suitable phase spaces)? Are the results of Markus-Meyer adaptable, and how? Note, however, that the results of Markus and Mayer are in contrast to the result of Oxtoby and Ulam that a generic set of conservative topological flows is ergodic.

Based on numerical simulations some researchers suggest that for large perturbations of an integrable Hamiltonian the system can be nonergodic due to a portion of phase space of very small Lebesgue measure, that is, the orbits of a large set of initial conditions would be dense in $\Gamma \setminus \Lambda$, and Λ with small Lebesgue measure, so that for most initial conditions time averages would result in values near the microcanonical average, and relation would hold for practical purposes. This is an attractive argument after Markus-Meyer generic result, but a rigorous approach is still missing. Instead of large perturbations one could think of small perturbations but with large number of particles, which is again related to Fermi-Pasta-Ulam work; even here the resulting picture indicates a difficult problem as illustrated in numerical studies.

Another possibility is mentioned by Dobrushin: for a large number of particles it is possible that a lot of ergodic components are present and so intermixed that in each finite subvolume it becomes difficult to distinguish those components.

Fluctuations in Different Ensembles

Fluctuations in Canonical Ensemble

(i) Relative root mean square fluctuation in energy in canonical ensemble (equivalence of canonical ensemble with microcanonical ensemble).

Let us take the derivative of average energy in canonical ensemble with respect to β.

$$\frac{\partial\langle E\rangle}{\partial\beta}=\frac{\partial\left\{\frac{\sum_r E_r e^{-\beta Er}}{\sum_r e^{-\beta Er}}\right\}}{\partial\beta}=-\left\{\frac{\sum_r E_r^2 e^{-\beta Er}}{\sum_r e^{-\beta Er}}\right\}+\left\{\frac{\left(\sum_r E_r e^{-\beta Er}\right)^2}{\left(\sum_r e^{-\beta Er}\right)^2}\right\}$$

$$\text{var iance}=\left\langle(\Delta E)^2\right\rangle=-\frac{\partial\langle E\rangle}{\partial\beta}=\left\langle E^2\right\rangle\langle E\rangle^2=k_B T^2\frac{\partial E_r}{\partial T}=k_B T^2 C_v$$

Let us now calculate relative root mean square fluctuation,

$$\frac{\sqrt{\langle\Delta E\rangle^2}}{\langle E\rangle}=\frac{\sqrt{k_B T^2 C_v}}{\langle E\rangle}$$

Now to estimate this relative root mean square fluctuations, let us recall for an ideal gas $\langle E\rangle=3Nk_BT$ and $C_v=3Nk_B$ where N is the number of particles in the system, therefore,

$$\frac{\sqrt{\langle\Delta E\rangle^2}}{\langle E\rangle}=\frac{\sqrt{\langle k_B T\rangle^2 C_v}}{\langle E\rangle}=\frac{\sqrt{k_B T^2 3Nk_B}}{3Nk_B T}\sim\frac{1}{\sqrt{N}}$$

Thus relative root mean square fluctuation is of the order of $\frac{1}{\sqrt{N}}$. For 1 mole of an ideal gas number of particles N is of the order of 10^{23} hence variance is of the order of 10^{-12} of the mean energy $\langle E\rangle$ a very small spread about the mean energy. Leading to an important conclusion that it is the same as we find in micro canonical ensemble i.e. the two ensembles give the same thermo-dynamical result.

Fluctuations in Grand Canonical Ensemble

(ii) Relative root mean square fluctuation in number of particles and energy in grand canonical ensemble (equivalence of grand canonical ensemble with other ensembles).

a) Relative root mean square fluctuation in number of particles:

Let us take the derivative of average number of particles with respect to α, and noting that $\alpha=-\frac{\mu}{k_B T}=-\mu\beta,$

$$\frac{\partial\langle N\rangle}{\partial\alpha}=\frac{\partial\left\{\frac{\sum_{sr} N_r e^{-\beta E_s-\alpha N_r}}{\sum_{sr} e^{-\beta E_s-\alpha N_r}}\right\}}{\partial\alpha}=-\left\{\frac{\sum_{sr} N_r^2 e^{-\beta E_s-\alpha N_r}}{\sum_{sr} e^{-\beta E_s-\alpha N_r}}\right\}+\left\{\frac{\left(\sum_{sr} N_r e^{-\beta E_s-\alpha N_r}\right)^2}{\left(\sum_{sr} N_s e^{-\beta E_s-\alpha N_r}\right)^2}\right\}$$

$$\text{var iance}=\langle\Delta N\rangle^2=-\frac{\partial\langle N\rangle}{\partial\alpha}=\left\langle N^2\right\rangle\langle N\rangle^2=k_B T\left(\frac{\partial N}{\partial\mu}\right)_{T,V}$$

Let us now calculate relative root mean square fluctuation,

$$\frac{\sqrt{(\Delta N)^2}}{\langle N\rangle}=\frac{\sqrt{k_B T\left(\frac{\partial N}{\partial \mu}\right)_{T,V}}}{\langle N\rangle}$$

Now to estimate this relative root mean square fluctuations, we can observe that in equation below numerator is of the order of $\sqrt{N}$ and denominator of the order of N, the number of particles in the system, therefore,

$$\frac{\sqrt{(\Delta N)^2}}{\langle N\rangle}=\frac{\sqrt{k_B T\left(\frac{\partial N}{\partial \mu}\right)_{T,V}}}{\langle N\rangle}\sim\frac{1}{\sqrt{N}}$$

Thus relative root mean square fluctuation is of the order of $\frac{1}{\sqrt{N}}$. For 1 mole of an ideal gas number of particles N is of 10^{23} the order of hence variance is of the order of 10^{-12} of the mean number of particles a very small spread about the mean number of particles. Leading to an important conclusion that it is the same as we find in micro canonical ensemble i.e. the two ensembles the grand canonical and microcanonical give the same thermos-dynamical result.

We can also express equation above in terms of isothermal compressibility, $k_T=-\frac{1}{V}\left(\frac{\partial V}{\partial P}\right)_T$ Recall Gibbs-Duhem relation for a thermodynamic system according to which,

$$Nd\mu=-SdT+VdP$$

Therefore,

$$\left(\frac{\partial \mu}{\partial N}\right)_{T,V}=\frac{V}{N}\left(\frac{\partial P}{\partial N}\right)_{T,V}$$

and

$$\left(\frac{\partial \mu}{\partial V}\right)_{T,N}=\frac{V}{N}\left(\frac{\partial P}{\partial V}\right)_{T,N}$$

Also in terms of Helmholtz free energy, F,

$$dF=-SdT-PdV+\mu dN$$

We have,

$$P=-\left(\frac{\partial F}{\partial V}\right)_{T,N} \text{ and } \mu=\left(\frac{\partial F}{\partial N}\right)_{T,V}$$

From above equation it follows that,

$$\left(\frac{\partial P}{\partial N}\right)_{T,V}=-\left(\frac{\partial F}{\partial N\partial V}\right)_T \text{ and } \left(\frac{\partial \mu}{\partial V}\right)_{T,N}=\left(\frac{\partial F}{\partial V\partial N}\right)_T$$

Therefore,

$$\left(\frac{\partial P}{\partial N}\right)_{T,V}=-\left(\frac{\partial \mu}{\partial V}\right)_{T,N}$$

Using equations $\left(\frac{\partial \mu}{\partial N}\right)_{T,V}=\frac{V}{N}\left(\frac{\partial P}{\partial N}\right)_{T,V}$ and $\left(\frac{\partial \mu}{\partial V}\right)_{T,N}=\frac{V}{N}\left(\frac{\partial P}{\partial V}\right)_{T,N}$ along with $\left(\frac{\partial P}{\partial N}\right)_{T,V}=-\left(\frac{\partial \mu}{\partial V}\right)_{T,N}$ we have,

$$\left(\frac{\partial \mu}{\partial N}\right)_{T,V}=\frac{V}{N}\left(\frac{\partial P}{\partial N}\right)_{T.V}=-\frac{V}{N}\left(\frac{\partial \mu}{\partial V}\right)_{T,N}=-\frac{V^2}{N^2}\left(\frac{\partial P}{\partial V}\right)_{T,N}$$

Using above equation in ($\frac{\sqrt{(\Delta N)^2}}{\langle N\rangle}=\frac{\sqrt{k_BT\left(\frac{\partial N}{\partial \mu}\right)_{T,V}}}{\langle N\rangle}\sim\frac{1}{\sqrt{N}}$), we have

$$\frac{\sqrt{(\Delta N)^2}}{\langle N\rangle}=\frac{\sqrt{\frac{k_BTN^2}{V}\left(-\frac{1}{V}\left(\frac{\partial V}{\partial P}\right)_{T,N}\right)}}{\langle N\rangle}=\frac{\sqrt{\frac{k_BTN^2}{V}k_T}}{\langle N\rangle}=\sqrt{\frac{k_BT}{V}k_T}$$

(b) Relative root mean square fluctuation in energy:

Let us take the derivative of average energy in canonical ensemble with respect to β and noting that $\alpha=-\frac{\mu}{k_BT}=-\mu\beta$.

$$\frac{\partial\langle E\rangle}{\partial\beta}=\frac{\partial\left\{\frac{\sum_{sr}E_se^{-\beta E_s-\alpha N_r}}{\sum_{sr}e^{-\beta E_s-\alpha N_r}}\right\}}{\partial\beta}=-\left\{\frac{\sum_{sr}E_s^2e^{-\beta E_s-\alpha N_r}}{\sum_{sr}e^{-\beta E_s-\alpha N_r}}\right\}+\left\{\frac{\left(\sum_{sr}E_se^{-\beta E_s-\alpha N_r}\right)^2}{\left(\sum_{sr}E_se^{-\beta E_s-\alpha N_r}\right)^2}\right\}$$

$$\text{variance}=\langle\Delta E\rangle^2=-\frac{\partial\langle E\rangle}{\partial\beta}=\langle E^2\rangle\langle E\rangle^2=k_BT\left(\frac{\partial E}{\partial T}\right)_{\alpha,V}$$

Before we proceed further let us carefully look at $\left(\frac{\partial E}{\partial T}\right)\alpha,V$, since N is not constant it is a contribution should be there because of fluctuation in N, therefore it becomes,

$$\left(\frac{\partial E}{\partial T}\right)_{\alpha,V}=\left(\frac{\partial E}{\partial T}\right)_{N.V}+\left(\frac{\partial E}{\partial N}\right)_{T,V}\left(\frac{\partial N}{\partial T}\right)_{\alpha,V}$$

Here first term is the specific heat at constant volume, C_V as was in canonical ensemble and the seond term takes care of the particle number fluctuation.

Recalling

$$\langle E\rangle = -\left[\frac{\partial q(\mu,V,T)}{\partial\beta}\right]_{V\alpha}$$

$$\langle N\rangle k_B T\left[\frac{\partial q(\mu,V,T)}{\partial\mu}\right]_{V,T} = -\left[\frac{\partial q(\mu,V,T)}{\partial\alpha}\right]_{V,T}$$

We have,

$$\left(\frac{\partial\langle E\rangle}{\partial\alpha}\right)_{V,T} = -\left[\frac{\partial^2 q(u,V,T)}{\partial\alpha\partial\beta}\right]_{V,f} = \left(\frac{\partial\langle N\rangle}{\partial\beta}\right)_{\alpha,V}$$

Noting that $\left(\frac{\partial\langle E\rangle}{\partial\alpha}\right)_{V,T} = -k_B T\left(\frac{\partial\langle E\rangle}{\partial\mu}\right)_{V,T}$ and $\left(\frac{\partial\langle N\rangle}{\partial\beta}\right)_{\alpha,V} = -T^2 k_B\left(\frac{\partial\langle N\rangle}{\partial T}\right)_{\mu V}$ equation above gives

$$\frac{1}{T}\left(\frac{\partial\langle E\rangle}{\partial\mu}\right)_{V,T} = \left(\frac{\partial\langle N\rangle}{\partial T}\right)\mu, V.$$

And equation $\left(\frac{\partial E}{\partial T}\right)_{\alpha,V} = \left(\frac{\partial E}{\partial T}\right)_{N.V} + \left(\frac{\partial E}{\partial N}\right)_{T,V}\left(\frac{\partial N}{\partial T}\right)_{\alpha,V}$ becomes,

$$\left(\frac{\partial E}{\partial T}\right)_{\alpha,V} = \left(\frac{\partial E}{\partial T}\right)_{N.V} + \frac{1}{T}\left(\frac{\partial E}{\partial N}\right)_{T,V}\left(\frac{\partial\langle E\rangle}{\partial T}\right)_{V,T}$$

Putting above equation in ($variance = \langle\Delta E\rangle^2 = -\frac{\partial\langle E\rangle}{\partial\beta} = \langle E^2\rangle\langle E\rangle^2 = k_B T\left(\frac{\partial E}{\partial T}\right)_{\alpha,v}$) we get

$$\langle\Delta E\rangle^2 = k_B T^2\left(\frac{\partial E}{\partial T}\right)_{\alpha,v} = k_B T^2\left(\frac{\partial E}{\partial T}\right)_{N,V} + k_B T\left(\frac{\partial E}{\partial N}\right)_{T,V}\left(\frac{\partial\langle E\rangle}{\partial\mu}\right)_{V,T}$$

Recalling $\left(\frac{\partial E}{\partial T}\right)_{N,V} = C_V$ and $\langle\Delta N\rangle^2 = k_B T\left(\frac{\partial N}{\partial\mu}\right)_{T,V} = k_B T\left(\frac{\partial N}{\partial E}\right)_{T,V}\left(\frac{\partial E}{\partial\mu}\right)_{T,V}$

Equation above can be written as:

$$\langle\Delta E\rangle^2 = k_B T^2\left(\frac{\partial E}{\partial T}\right)_{\alpha,v} = k_B T^2 C_V + \left\{\left(\frac{\partial E}{\partial N}\right)_{T,V}\right\}^2\langle\Delta N\rangle^2$$

Thus in grand canonical variance is made up of two parts, first term is the canonical contribution and the second part comes because of particle number fluctuation. The seond term becomes important during phase transitions when $\langle\Delta N\rangle^2$ becomes exceptionally large.

Thus relative fluctuation in energy in Grand Canonical Ensemble becomes,

$$\frac{\sqrt{(\Delta E)^2}}{\langle E\rangle}=\frac{\sqrt{k_B T N^2\left(\frac{\partial E}{\partial T}\right)_{\alpha.V}}}{3Nk_B T}+\frac{\sqrt{k_B T^2 C_V+\left\{\left(\frac{\partial E}{\partial N}\right)_{T,V}\right\}^2}}{3k_B T}=\sqrt{\frac{\langle \Delta N\rangle^2}{N}}.$$

Importance of Fluctuations

We had set upon establishing equivalence of three types of ensembles. We found that in each case relative fluctuations vary as $\frac{1}{\sqrt{N}}$ and for a macroscopic system in thermodynamic limit $(N \to \infty)$, relative fluctuations vanish. However, there are situations in which, fluctuations can be very large and this result may not be true. This happens in the case of phase transitions where there are large fluctuations in number density,

$$\frac{\sqrt{(\Delta N)^2}}{\langle N\rangle}=\sqrt{\frac{k_B T}{V}k_T}$$

because isothermal compressibility of the system there becomes exceptionally large. Critical Opalescence is one such phenomenon which happens near critical point in binary liquid mixtures when visible light gets scattered very strongly and clear mixture turns turbid.

Bose-Einstein, Fermi- Dirac Distribution and Maxwell Boltzmann Distributions from Microcanonical Ensemble

Now to further see the equivalence between different ensembles, let us try to derive Bose-Einstein distribution and Fermi-Dirac Distributions from microcanonical ensemble. We derived this result earlier in Grand Canonical Ensemble. It shall be interesting to note that in spite of different surroundings provided to the system, the same results get reproduced. The system we have chosen is a gaseous system of fixed N number of non-interacting indistinguishable particles enclosed in a volume V with fixed total energy E.

We also know that in thermodynamic limit in a macroscopic system, the single particle energy Eigen values form a closely spaced set of energy levels. In this situation one can divide the spectrum in to a collection of cells, with each cell containing a cluster of levels with each cluster I having an average energy ϵ_i and and an arbitrary number of energy levels $g_i \gg 1$. In a particular state of the system first cluster may contain n_1 number of particles, second cluster may contain n_2 number of particles and so on such that the distribution set of particles $\{n_i\}$ should satisfy the conditions as required by the microcanonical ensemble.

$$\sum_i n_i = N$$

and

$$\sum_i n_i \epsilon_i = E$$

Pictorially situation is as shown in figure below:

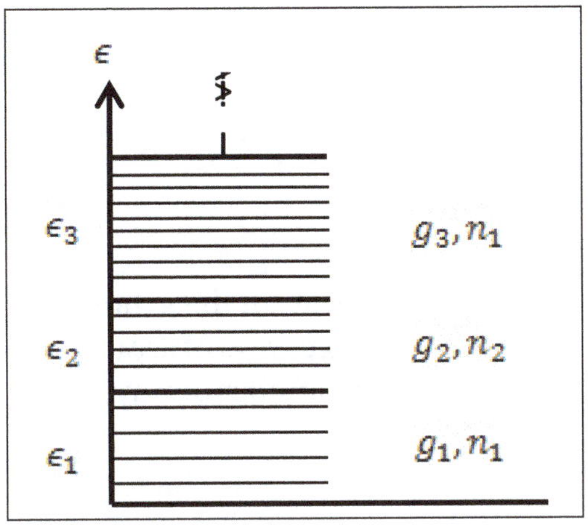

For a particular set of values of number of particles, i.e. occupation number set $\{n_i\}$ distributed in a cell the number of microstates is say $\Omega\{n_i\}$, such that the total number of microstates $\Omega(N,V,E)$ is given by,

$$\Omega(N,V,E)=\sum_{\{n_i\}}\Omega\{n_i\}$$

Figure Clustering of single particle energy levels.

The $\Omega\{n_i\}$ can be further expressed in terms of number of possible ways in which n_i particles can be distributed in each of the ith cell denoted by, such that,

$$\Omega\{n_i\}=\prod_i \omega(i)$$

Where $\omega(i)$ depends on the type of particles, whether they are fermions or bosons or classical.

Calculation Number of Microstates in Different Statistics

Bosons

In the case of bosons, in which any number of particles can be put in an energy level, problem is similar to putting n_i number of particles in g_i number of cells, which can be achieved by placing n_i number of particles on a straight line with in two walls and permuting g_i -1 number of partitions to create g_i number of cells or energy levels. This amounts to shuffling g_i -1 number of partitions within the walls to enclose any number of particles in a cell. The total number of particles and partition walls is n_i+g_i-1, and this can be achieved by placing g_i-1 number of partitions at any one of the locations n_i+g_i-1.

The number required is,

$$\omega_{BE}(i)=\frac{(n_i+g_i-1)!}{n_i!(g_i-1)!}$$

Thus using equation $\Omega\{n_i\}=\prod_i \omega(i)$ the total number of microstates for a given distribution $\Omega\{n_i\}$ is given by,

$$\Omega_{BE}\{n_i\}=\prod_i \frac{(n_i+g_i-1)!}{n_i!(g_i-1)!}$$

Fermions

In the case of fermions, n_i, the number of particles in the g_i number of energy levels or cells is such that $n_i \le g_i$. Thus $\omega_{FD}(i)$ is the number of possible ways of arranging n_i particles in g_i states so that no more than one particle in each state is equal to number of ways of selecting n_i states out of g_i states.

$$\omega_{FD}(i)\frac{g_i!}{n_i!(g_i-n_i)!}$$

Thus using equation $\Omega\{n_i\}=\prod_i \omega(i)$ the total number of microstates for the given distribution $\Omega\{n_i\}$ is given by,

$$\Omega_{FD}\{n_i\}=\prod_i \frac{g_i!}{n_i!(g_i-n_i)!}$$

Maxwell-Boltzmann Case

In this case particles are distinguishable, so that any of the n_i particles may be put in g_i energy levels leading to distinct states. So the first particle can be put in g_i ways, second particle can also be put in g_i ways and so on so that the number of distinct ways in putting n_i particles together in g_i states is $g_i^{n_i}$.

Furthermore, distribution set $\{n_i\}$ can be obtained by selecting out of N number of total particles n_1 number of particles in first cluster, n_2 number of particles in the second cluster and so on this distribution itself can be obtained in,

$$\frac{N!}{n_1!n_2!n_3!\ldots}$$

Combining results ($g_i^{n_i}$) and ($\frac{N!}{n_1!n_2!n_3!\ldots}$),

$$\Omega_{MB}\{n_i\}=\prod_i \frac{N!\,g_i^{n_i}}{n_i!}$$

However this result needs to be corrected to take care of Gibbs paradox by dividing by N! factor.

Thus finally the number of microstates in Maxwell-Boltzmann distribution is,

$$\Omega_{MB}\{n_i\}=\prod_i \frac{g_i^{n_i}}{n_i!}$$

For each of the cases for the three statistics we are now ready to go ahead and calculate using equations $\Omega_{BE}\{n_i\}=\prod_i \frac{(n_i+g_i-1)!}{n_i!(g_i-1)!}$, $\Omega_{FD}\{n_i\}=\prod_i \frac{g_i!}{n_i!(g_i-n_i)!}$ and $\Omega_{MB}\{n_i\}=\prod_i \frac{g_i^{n_i}}{n_i!}$ respectively for Bose-Einstein statistics, Fermi-Dirac Statistics and Maxwell Boltzmann statistics respective distribution functions.

Calculation of Most Probable Distribution in Different Statistics

One common link in each of the cases is that we are interested in the distribution which corresponds to maximum number of microstates Ω given by equation $\Omega(N,V,E)=\sum_{\{n_i\}}\Omega\{n_i\}$, since this corresponds to the most probable distribution, making entropy S of the system maximum. Since entropy is,

$$S(N,V,E)=k_B \text{ In } \Omega(N,V,E)=k_B \text{ In}\sum_{\{n_i\}}\Omega\{n_i\}$$

it implies maximizing $\text{In}\sum_{\{n_i\}}\Omega\{n_i\}$ .Here, there is an important observation regarding this sum in the argument of log term, which is the fact that the sum $\sum_{\{n_i\}}\Omega\{n_i\}$ can be approximated by keeping only the largest term in the summation provided n_i and g_i are large that is the term $\Omega\{n_i\}_{max}$, where $\{n_i\}_{max}$, defines the most probable distribution provided there is no violation of the constraints ($\sum_i n_i=N$) and ($\sum_i n_i\in_i=E$), keeping total energy and total number of particles fixed. So we can once again apply lagrange's method of undetermined multipliers, so that condition for knowing $\{n_i\}_{max}$, becomes,

$$\delta \text{ In}\Omega\{n_i\}-\left[\alpha\sum_i\delta n_i+\beta\sum_i\in_i\delta n_i\right]=0$$

Or,

$$\delta \text{ In}\left(\prod_i\omega_{dist}(n_i)\right)-\left[\alpha\sum_i\delta n_i+\beta\sum_i\in_i\delta n_i\right]=0$$

Or,

$$\delta\sum_i \text{In } \omega_{dist}(n_i)-[\alpha\delta n_i+\beta\in_i\delta n_i]=0$$

Now let us using Stirling's approximation write down for each of the statistics $\text{In } \omega_{dist}(n_i)$:

Case: Bose Einstein Statistics

$$\sum_i \text{In } \omega_{SE}(n_i)=\sum_i \text{In}\frac{(n_i+g_i-1)}{(n_i!+g_i-1)!}$$
$$=\sum_i\left[(n_i+g_i-1)\text{ In }(n_i+g_i-1)-(n_i)\text{In}(n_i)-(g_i-1)\text{ In }(g_i-1)\right]$$

Or for $n_i\gg1$ and $g_i\gg1$,

$$\sum_i \text{In } \omega_{EE}(n_i) = \sum_i \left[n_i \text{ In}\left(1+\frac{g_i}{n_i}\right) + g_i \text{ In})\frac{n_i}{g_i}+1 \right]$$

Let us find variation of above equation with respect to n_i:

$$\sum_i \delta \text{ In } \omega_{BE}(n_i) = \sum_i \left[\delta n_i \text{ In}\left(1+\frac{g_i}{n_i}\right) + n_i\left(\frac{1}{n_i+g_i}\right)\delta n_i - n_i\left(\frac{1}{n_i}\right)\delta n_i + g_i\left(\frac{1}{n_i+g_i}\right)\delta n_i \right]$$

$$= \sum \text{In}\left(1+\frac{g_i}{n_i}\right)\delta n_i.$$

Case: Fermi-Dirac statistics:

$$\sum_i \text{In } \omega_{FD}(i) = \sum_i \text{In}\frac{g_i!}{n_i!(g_i-n_i)!} = \sum_i [g_i \text{ In } g_i - n_i \text{ In } n_i - (g_i-n_i)\text{In }(g_i-n_i)]$$

Or,

$$\sum_i \text{In } \omega_{FD}(i) = \sum_i \left[n_i \text{ In}\left(\frac{g_i}{n_i}-1\right) - g_i \text{ In}\left(1-\frac{n_i}{g_i}\right)\right]$$

Let us find variation of above equation with respect to n_i

$$\delta\sum_i \text{In } \omega_{FD}(i) = \sum_i \delta n_i \text{ In}\left(\frac{g_i}{n_i}-1\right) - n_i\frac{1}{g_i-n_i}\delta n_i + n_i\frac{1}{n_i}\delta n_i + g_i\left(\frac{1}{g_i-n_i}\right)\delta n_i$$

$$= \sum_i \delta n_i \text{ In}\left(\frac{g_i}{n_i}-1\right)$$

Case: Maxwell-Boltzmann Statistics:

$$\sum_i \text{In } \omega_{MB}(i) \sum_i \text{In}\frac{g_i^{n_i}}{n_i!} = \sum_i [n_i \text{ In} g_i - n_i \text{ In } n_i + n_i]$$

Let us find variation of above equation with respect to n_i

$$\delta\sum_i \text{In } \omega_{MB}(i) = \sum_i \left[\delta n_i \text{ In } g_i - \delta n_i \text{ In } n_i - n_i\frac{1}{n_i}\delta n_i + \delta n_i \right] = \sum_i \left[\delta n_i \text{ In}\frac{g_i}{n_i} \right]$$

Equations, $= \sum \text{In}\left(1+\frac{g_i}{n_i}\right)\delta n_i$, $= \sum \delta n \text{ In}\left(\frac{}{} - 1\right)$ and above equation can be combined together in terms of a parameter η, such that,

$$\delta\sum_i \text{In } \omega_{dist}(i) = \sum_i \left[\delta n_i \text{ In}\left(\frac{g_i}{n_i}-\eta\right)\right]$$

Where,

$$\eta = \begin{cases} -1 \text{ for Bose} - \text{Einstein} \\ 1 \text{ for fermi Dirac,} \\ 0 \text{ for Maxwell} - \text{Boltzmann.} \end{cases}$$

Substituting ($\delta\sum_i \text{In}\ \omega_{dist}(i) = \sum_i \left[\delta n_i \text{ In}\left(\frac{g_i}{n_i} - \eta\right)\right]$) in equation $\delta\sum_i \text{In}\ \omega_{dist}(n_i) - [\alpha\delta n_i + \beta \in_i \delta n_i] = 0$ we have,

$$\sum_i \left[\text{In}\left(\frac{g_i}{n_i} - \eta\right) - [\alpha + \beta \in_i]\right]\delta n_i = 0$$

Since all δn_i are independent, for equation to be true, the co-efficients of each δn_i must vanish, i.e.

$$\left[\text{In}\left(\frac{g_i}{n_i} - \eta\right) - [\alpha + \beta \in_i]\right] = 0$$

Which implies,

$$\frac{g_i}{n_i} = e^{\alpha + \beta \in_i} + \eta$$

Or occupation number $\frac{n_i}{g_i}$ the most probable number of particles in a single energy level can be obtained as,

$$\frac{n_i}{g_i} = \frac{1}{e^{\alpha + \beta \in_i} + \eta}$$

Entropy and Equation of State

From equation $S(N,V,E) = k_B \text{ In } \Omega(N,V,E) = k_B \text{ In} \sum_{\{n_i\}} \Omega\{n_i\}$ entropy can be written for all these cases by combining the following three results into a single result in terms of earlier defined parameter η as,

$$\sum_i \text{In}\ \omega_{BE}(n_i) = \sum_i \left[n_i \text{ In}\left(1 + \frac{g_i}{n_i}\right) + g_i \text{ In}\left(\frac{n_i}{g_i} + 1\right)\right]$$

$$\sum_i \text{In}\ \omega_{FD}(i) = \sum_i \left[n_i \text{ In}\left(\frac{g_i -}{n_i} - 1\right) - g_i \text{ In}\left(1 - \frac{n_i}{g_i}\right)\right]$$

$$\sum_i \text{In}\ \omega_{MB}(i) = \sum_i \left[n_i \text{ In}\frac{g_i}{n_i} + n_i\right]$$

$$\sum_i \text{In}\ \omega_\eta(n_i) \sum_i \left[n_i \text{ In}\left(-\eta + \frac{g_i}{n_i}\right) - \frac{g_i}{\eta} \text{In}\left(-\eta\frac{n_i}{g_i} + 1\right)\right]$$

So that,

$$\frac{S}{k_B} \approx \sum_i \text{In } \omega_\eta(n_i) = \sum_i \left[n_i \text{ In}\left(-\eta + \frac{g_i}{n_i}\right) - \frac{g_i}{\eta} \text{In}\left(-\eta \frac{n_i}{g_i} + 1\right) \right]$$

Using relation ($\frac{n_i}{g_i} = \frac{1}{e^{\alpha+\beta\epsilon_i} + \eta}$) we have,

$$\frac{S}{k_B} \approx \sum_i \left[n_i(\alpha + \beta \in_i) + \frac{g_i}{\eta} \text{In}\left(\eta e^{-\alpha-\beta\epsilon_i} + 1\right) \right]$$

Using the constraints ($\sum_i n_i = N$) and ($\sum_i n_i \in_i = E$) for fixed number of total particles and fixed total energy we can rewrite equation $\frac{S}{k_B} \approx \sum_i \text{In } \omega_\eta(n_i) = \sum_i \left[n_i \text{ In}\left(-\eta + \frac{g_i}{n_i}\right) - \frac{g_i}{\eta} \text{In}\left(-\eta \frac{n_i}{g_i} + 1\right) \right]$ as,

$$\frac{S}{k_B} = (N\alpha + \beta E) + \sum_i \left[+ \frac{g_i}{\eta} \text{In}\left(\eta e^{-\alpha-\beta\epsilon_i} + 1\right) \right]$$

Or,

$$\sum_i \left[+ \frac{g_i}{\eta} \text{In } \left(\eta e^{-\alpha-\beta\epsilon_i} + 1\right) \right] = \frac{S}{k_B} - (N\alpha + \beta E)$$

The Lagrange's undetermined multipliers α and β can be interpreted as,

$$\alpha = -\frac{\mu}{k_B T} \text{ and } \beta = \frac{1}{k_B T}$$

So that right hand side of equation $\sum_i \left[+ \frac{g_i}{\eta} \text{In } \left(\eta e^{-\alpha-\beta\epsilon_i} + 1\right) \right] = \frac{S}{k_B} - (N\alpha + \beta E)$ can be written as,

$$\frac{S}{k_B} + \frac{N\mu}{k_B T} - \frac{E}{k_B T} = \frac{G - (E - TS)}{k_B T} = \frac{PV}{k_B T}$$

Or,

$$\frac{PV}{k_B T} = \sum_i \left[+ \frac{g_i}{\eta} \text{In}\left(\eta e^{-\alpha-\beta\epsilon_i} + 1\right) \right]$$

In Maxwell-Boltzmann case $\eta \to o$, equation above gives ideal equation of state:

$$PV = k_B T \sum_i \left[g_i e^{-\alpha-\beta\epsilon_i} \right] = k_B T \sum_i [n_i] = N k_B T$$

It is interesting to note here that g_i which is multiplicity (or degeneracy) of energy states are so close that summation in equation $\frac{PV}{k_BT}=\sum_i\left[+\frac{g_i}{\eta}\text{In}\left(\eta e^{-\alpha-\beta\epsilon_i}+1\right)\right]$ can converted into an integration over momentum space recalling that,

$$g_i=\frac{4\pi Vp^2dp}{h^3}$$

So that pressure can be obtained as,

$$P=\frac{4\pi k_BT}{\eta h^3}\int_o^\infty \text{In}\left(\eta e^{-\alpha-\beta\epsilon(p)}+1\right)p^2dp$$

This can be integrated by parts to get,

$$=\frac{4\pi k_BT}{\eta h^3}\left[\frac{p^3}{3}\text{In}\left(\eta e^{-\alpha-\beta\epsilon(p)}+1\right)\Big|_o^\infty+\int_o^\infty\frac{p^3}{3}\frac{\eta e^{-\alpha-\beta\epsilon(p)}}{\left(\eta e^{-\alpha-\beta\epsilon(p)}+1\right)}\beta\frac{d\epsilon(p)}{dp}dp\right]$$

First term in above equation vanishes at both the limits. So that,

$$P=\frac{4\pi}{3h^3}\int_o^\infty\frac{p^3}{\left(\eta+e^{\alpha+\beta\epsilon(p)}\right)}\left(p\frac{d\epsilon(p)}{dp}\right)dp$$

Recalling that,

$$N=\sum_i n_i=\sum_i\frac{g_i}{e^{\alpha+\beta\epsilon_i+\eta}}$$

$$=\frac{4\pi V}{h^3}\int_o^\infty\frac{p^2}{\left(\eta+e^{\alpha+\beta\epsilon(p)}\right)}dp=\int_o^\infty\frac{4\pi V}{h^3}\frac{p^2}{\left(\eta+e^{\alpha+\beta\epsilon(p)}\right)}dp=\int_o^\infty dN$$

Equation $P=\frac{4\pi}{3h^3}\int_o^\infty\frac{p^3}{\left(\eta+e^{\alpha+\beta\epsilon(p)}\right)}\left(p\frac{d\epsilon(p)}{dp}\right)dp$ can be written as,

$$P=\frac{1}{3V}\int_o^\infty dN\left(p\frac{d\epsilon(p)}{dp}\right)=\frac{N}{3V}\left\langle p\frac{d\epsilon(p)}{dp}\right\rangle$$

Noting that generally relation between ϵ and p is of the type $\epsilon\propto p^r$, $p\frac{d\epsilon(p)}{dp}=r\epsilon$ and hence P becomes,

$$P=\frac{rN}{3V}\langle\epsilon\rangle$$

For a classical ideal gas $\langle\epsilon\rangle=\frac{3}{2}k_BT$ and $r=2$, therefore, $PV=Nk_BT$ an ideal gas equation.

Interestingly this result applies to all the three statistics.

Quantum Ensemble Theory

Ensemble interpretations of quantum theory contend that the wave function describes an ensemble of identically prepared systems. They are thus in contrast to “orthodox” or “Copenhagen” interpretations, in which the wave function provides as complete a description as is possible of an individual system. A study of the relative merits of these two classes of interpretation, and in particular of the extent to which ensemble interpretations may avoid the well-known difficulties of orthodox interpretations. Ensemble interpretations of quantum theory have practically as long a pedigree as orthodox interpretations. If Bohr and Heisenberg may be regarded as the founders of the latter, Einstein was an early supporter of the view, which he maintained until the end of his life, that quantum theory represented a correct and complete statistical theory of ensembles, but not a theory of elementary processes. In other words he regarded the theory as incomplete, which was indeed the conclusion of the famous Einstein—Podolsky—Rosen (EPR) argument. In the early days of the quantum theory, ensemble interpretations had other strong advocates — Slater, Kemble, Popper, Langevin,and Margenau, but the very great influence of Bohr meant that, at the very least till the 1950s, such views remained peripheral. One should also mention the famous argument of von Neumann, which claimed to show that hidden-variable theories were impossible; in view of our remarks immediately above, it would be expected that acceptance of such arguments would at least dampen enthusiasm for ensemble interpretations.

Classically, state of the system is said to be specified once the position and momenta of all particles constituting the system gets specified. However, because of Heisenberg's Uncertainty principle, position and momenta cannot be specified simultaneously. In quantum mechanics, we deal with Hermitian operators which represent physically observable quantities with real eigen values. Some often encountered operators are $\hat{H}$, $\hat{L}$ and $\hat{S}$ representing respectively energy, angular momentum and spin angular momentum. Many physical quantities, like spin angular momentum, have no classical analogue. Each of the operators operates on a vector space the so called Hilbert space. The state of the system at a certain time t is specified by a state vector $|\psi(t)\rangle$. Time evolution of the state $|\ \ (\ )\rangle$ is governed via the Schrodinger equation,

$$i\hbar\frac{\partial|\psi(t)\rangle}{\partial t}=\hat{H}|\psi(t)\rangle$$

Knowing $|\psi(t)\rangle$ allows one to calculate all characteristics of the system equivalent to knowing the microstate of the system. The co-ordinate representation of the system is described by the wave function $\psi(q,t)=\langle q|\psi(t)\rangle$ an inner product of spatial and temporal function. Quantum mechanics is inherently probabilistic in nature such that $|\psi(q,t)|^2$ is the probability density. The probability that a system or a particle is located in a volume element dq around q is given by $|\psi(q,t)|^2\,dq$. Given a physical quantity, represented by an operator $\hat{o}$ say, as we operate $\hat{o}$ on a quantum state, one gets its eigen value $\langle\hat{o}\rangle$, which is the expectation value of $\hat{o}$ which is the average value of the operator obtained during the process of measurement given by,

$$\langle\hat{o}\rangle=\frac{\int\psi^*(q,t)\hat{o}\psi(q,t)d\tau}{\int\psi^*(q,t)\psi^*(q,t)d\tau}$$

Where $d\tau$ is the volume element of the co-ordinate space. If $\psi(q,t)$ is normalized than,

$$\int \psi^*(q,t)\psi(q,t)d\tau = \int |\psi(q,t)|^2 d\tau = 1$$

For an operator $\hat{o}$ there can be a set of eigen values $\{o_n\}$ with a corresponding set of orthonormal eigen functions $\{\phi_n(q)\}$ forming a basis set, such that $\hat{o}\phi_n(q) = o_n\phi_n(q)$. The eigen function $\psi(q,t)$ can then be expanded in terms of this basis set as a result of superposition of basis eigen functions as,

$$\psi(q,t) = \sum_n a_n(t)\phi_n(q)$$

The physical significance of the coefficients lies in the fact that $|a_n(t)|^2$ gives the probability of getting in a measurement o_n corresponding to the eigen state $\phi_n(q)$ out of the set of eigen function $\{\phi_n(q)\}$, where $a_n(t)$ is,

$$a_n(t) = \int \phi_n^*(q)\psi(q,t)d\tau$$

Going to quantum statistical ensemble involves incorporation of quantum mechanical averaging besides statistical averaging leading to the so called double averaging.

Quantum Mechanical Formulation of Statistical Mechanics: Density Matrix

With the brief recapitulation of basic concepts of quantum mechanics given in the introduction we are now ready to apply these to a macroscopic system consisting of a large number of particles by using the notion of an ensemble. So we take an ensemble of $\mathcal{N}$ identical systems, where $\mathcal{N} >> 1$. Since these systems are identical they have a common Hamiltonian $\hat{H}$. The state of the various systems making the ensemble is represented by the wave functions $\psi(q,t)$, where $q = \{q_i\}$ denote the position coordinates relevant to the system being considered. Therefore, $\psi(q,t)$ describes an equivalent of a phase point in phase space of a classical system.

Let us focus on the kth system of the ensemble. Let at a given time t the kth system of the ensemble happens to be in a normalized state $\psi^k(q,t)$, where k=1, 2, 3, ... $\mathcal{N}$. The time evolution of this state function $\psi^k(q,t)$ is given by the time dependent Schrodinger equation,

$$i\hbar\frac{\partial\psi^k(t)}{\partial t} = \hat{H}\psi^k(t)$$

Where $\psi^k(t)$ can be expanded in terms of a set of orthonormal functions $\{\phi_n(q)\}$ as described in equation $\psi(q,t) = \sum_n a_n(t)\phi_n(q)$ in each system.

$$\psi^k(t) = \sum_n a_n^k(t)\phi_n(q)$$

Where,

$$a_n^k(t) = \int \phi_n^*(q)\psi^k(q,t)d\tau$$

Here $a_n^k(t)$ are the probability amplitudes of the system k to be in the state $\phi_n(q)$ and $|a_n^k(t)|^2$ is the probability amplitudes for the various systems of the ensemble in state $\phi_n(q)$. Sum of these probability amplitudes must be equal to 1.

$$\sum_n \left|a_n^k(t)\right|^2 = 1$$

Equation $i\hbar\frac{\partial\psi^k(t)}{\partial t} = \hat{H}\psi^k(t)$ can be written in terms of the coefficients $a_n^k(t)$ so that time variation of these coefficients is given by,

$$i\hbar\frac{\partial a_n^k(t)}{\partial t} = i\hbar\int\phi_n^*(q)\frac{\partial\psi^k(q,t)}{\partial t}d\tau$$

$$= \int\phi_n^*(q)\hat{H}\psi^k(t)d\tau$$

$$= \int\phi_n^*(q)\hat{H}\sum_n a_m^k(t)\phi_m(q)d\tau = \sum_m a_m^k(t)H_{mn}$$

Where,

$$H_{mn} = \int\phi_n^*(q)\hat{H}\phi_m(q)d\tau$$

Now let us define the density matrix by means of matrix elements,

$$\rho_{mn}(t) = \frac{1}{\mathcal{N}}\sum_{k=1}^{\mathcal{N}} a_m^k(t)a_n^{k*}(t)$$

Since sum here is over all the systems in the ensemble divided by number of systems in the ensemble, it gives an ensemble average of the quantity $a_m^k(t)a_n^{k*}(t)$. $m = n$ in equation above corresponds to the diagonal element $\rho_{nn}(t)$ which is the ensemble average of the probability density $\left|a_n(t)\right|^2$ which itself is a quantum average. So density matrix involves two averages, first one being a quantum mechanical average followed by an ensemble average.

So $\rho_{nn}(t)$ represents the probability that a system chosen at random from among the members of the ensemble shall be in the state ϕ_n at time t. Using equation $\sum_n \left|a_n^k(t)\right|^2 = 1$ we have,

$$\sum_n \rho_{nn}(t) = \frac{1}{\mathcal{N}}\sum_{k=1}^{\mathcal{N}}\sum_n a_n^k(t)a_n^{k*}(t) = \frac{1}{\mathcal{N}}\sum_{k=1}^{\mathcal{N}}\sum_n \left|a_n^k(t)\right|^2 = 1$$

Let us now express the ensemble average of an operator $\hat{G}$,

$$\langle G\rangle = \frac{1}{\mathcal{N}}\sum_{k=1}^{\mathcal{N}}\int\psi^{k*}\hat{G}\psi^k d\tau$$

This can further be written using $\psi^k(t) = \sum_n a_n^k(t)\phi_n(q)$ as,

$$\langle G\rangle = \frac{1}{\mathcal{N}}\sum_{k=1}^{\mathcal{N}}\left[\sum_{m,n} a_n^{*k}(t)a_m^k(t)G_{mn}\right]$$

Where,

$$G_{mn} = \int \phi_n^*(q)\hat{G}\phi_m(q)d\tau$$

Equation $\langle G\rangle = \frac{1}{\mathcal{N}}\sum_{k=1}^{\mathcal{N}}\left[\sum_{m.n} a_n^{*k}(t)a_m^k(t)G_{mn}\right]$ can be written as,

$$\langle G\rangle = \sum_{m.n}\left[\frac{1}{\mathcal{N}}\sum_{k=1}^{\mathcal{N}} a_n^{*k}(t)a_m^k(t)\right]G_{mn}$$

Or

$$\langle G\rangle = \sum_{m.n}\rho_{nm}G_{mn} = \sum_{m}\left(\hat{\rho}\hat{G}\right)_{mm} = \mathrm{Tr}\left(\hat{\rho}\hat{G}\right)$$

Note if $\hat{G} = \hat{1}$

Then,

$$\mathrm{Tr}\left(\hat{\rho}\right) = 1$$

In general if ψ^k are not normalized,

$$\langle G\rangle = \frac{\mathrm{Tr}\left(\hat{\rho}\hat{G}\right)}{\mathrm{Tr}\left(\hat{\rho}\right)}$$

It is obvious from the equation above knowledge of density matrix allows us to calculate mean value of any physical property of the system. Density matrix is the quantum analogue of the density function of the classical system.

Quantum Liouville's Theorem

Liouville's Theorem in quantum statistics is the quantum mechanical analogue of the Liouville's Theorem in classical statistics. It provides us information about the time dependence of density matrix. For this we need to determine the equation of motion of the density matrix. As a matter of convenience here after we well express derivative of a variable with respect to time by putting a dot over the variable, i.e. $\dot{\rho} = \frac{\partial p}{\partial t}$. Therefore, from equation $\rho_{mn}(t) = \frac{1}{\mathcal{N}}\sum_{k=1}^{\mathcal{N}} a_m^k(t)a_n^{k*}(t)$ we have,

$$i\hbar\dot{\rho}_{mn}(t) = \frac{1}{\mathcal{N}}\sum_{k=1}^{\mathcal{N}} i\hbar\left[\dot{a}_m^k(t)a_n^{k*}(t) + a_m^k(t)\dot{a}_n^{k*}(t)\right]$$

From equation $\sum_{n}\left|a_n^k(t)\right|^2 = 1$ it follows that,

$$i\hbar\dot{\rho}_{mn}(t) = \frac{1}{\mathcal{N}}\sum_{k=1}^{\mathcal{N}}\left[\left\{\sum_{l} a_l^k(t)H_{ml}\right\}a_n^{k*}(t) - a_m^k(t)\left\{\sum_{l} H^*_{nl}a_l^{*k}(t)\right\}\right]$$

Note the negative sign on the left hand side which arises because of complex conjugation procedure. Or equation above can be written as,

$$i\hbar\dot{\rho}_{mn}(t)=\left[\left\{\sum_{l}H_{ml}\rho_{ln}\right\}-\left\{\sum_{l}\rho_{ml}H^{*}_{nl}\right\}\right]$$

Since $\hat{H}$ is a Hermitian operator $H^{*}_{ln}=H_{ln}$

Therefore,

$$i\hbar\dot{\rho}_{mn}(t)=\left[\sum_{l}[H_{ml}\rho_{ln}-\rho_{ml}H_{ln}]\right]=\left(\hat{H}\hat{\rho}-\hat{\rho}\hat{H}\right)_{mn}$$

Or this equation can be written as,

$$i\hbar\dot{\rho}(t)=\left[\hat{H},\hat{\rho}\right]_{-}$$

If we compare this equation with classical Liouville's theorem, we will notice that Poisson bracket has been replaced $\frac{\left[\hat{H},\hat{\rho}\right]_{-}}{i\hbar}$.

Now coming back to the requirement for the system to be in equilibrium, the corresponding ensemble must be stationary, i.e. $\dot{\rho}_{mn}(t)=0$. For this condition to be true, we must have,

- $\hat{\rho}$ an explicit function of $\hat{H}$ i.e. $\rho(\hat{H})$, which implies that $\rho(\hat{H})$ commutes with $\hat{H}$.
- $\hat{H}$ should not be an explicit function of time t i.e. $\dot{\hat{H}}=0$.

Furthermore, if basis function set $\{\phi_n\}$ happens to be a set of eigen functions of $\hat{H}$ then the matrices defined by the elements H_{mn} and ρ_{mn} must be diagonal i.e.

$$H_{mn}=E_n\delta_{mn} \text{ and } \rho_{mn}=\rho_n\delta_{mn}$$

If the basis set happens to be other than basis set of the Hamiltonian, the matrices corresponding to $\hat{H}$ and $\hat{\rho}$ will not be diagonal, but it shall be symmetric i.e. $H_{mn}=H_{mn}$ and $\rho_{mn}=\rho_{nm}, P_{mn}=P_{nm}$, this is desirable for the system to go from state m $\rightarrow$ n must be the same as going from state n $\rightarrow$ m, satisfying the principle of detailed balancing.

Density Matrix and Various Ensembles

In the following we derive density matrix in various ensembles so that the result embodied in equation $\langle G\rangle=\frac{Tr(\hat{\rho}\hat{G})}{Tr(\hat{\rho})}$ may be applied to different systems in different surrounding environments, namely when system is completely isolated (micro canonical ensemble) from the surroundings, system is allowed to exchange energy with the surroundings (canonical ensermble) and when the system can exchange both energy and particles with the surroundings (Grand canonical ensemble).

The Microcanonical Ensemble

In micro canonical ensemble an isolated system in equilibrium is specified by constant values of number of particles N, constant volume V and energy lying in a narrow interval around $E:\left(E-\frac{1}{2}\Delta, E+\frac{1}{2}\Delta\right)$, with $\Delta << E$. Let the number of distinct microstates accessible to the system be Ω, which is to be computed quantum mechanically. Then according to postulate of equal a priori probability for each microstate is $\frac{1}{\Omega}$, then the density matrix in the energy representation, which has to be in diagonal form is,

$$\rho_{mn} = \rho_n \delta_{mn}$$

Where

$$\rho_n = \begin{cases} \frac{1}{\Omega} \text{for each accessible microstate} \\ 0 \text{ for all other micro states} \end{cases}$$

One can see that the normalization condition is clearly satisfied.

$$\sum_{mn} \rho_{mn} = \sum_{mn} \rho_n \delta_{mn} = \sum_{n} \rho_n = \frac{1}{\Omega}\left(\sum_{n} 1\right) = \frac{1}{\Omega}\Omega = 1$$

This result is true in representations other than energy representation considered here. There are two possible cases for the value of Ω, $\Omega = 1$ and $\Omega > 1$.

When Ω = 1, then the member systems of the ensemble are in the same state and hence there is one diagonal element $\rho_{mn} = \rho = 1$ and $\rho^2 = \rho$. Then according to Boltzmann formula entropy S = ln Ω, gives S = 0. The system is said to be in pure state.

When $\Omega > 1$, the system is said to be in mixed state. Ω for a macroscopic system is going to be very large number. In energy representation the density matrix is still given by equations $\rho_{mn} = \rho_n \delta_{mn}$

and $\rho_n = \begin{cases} \frac{1}{\Omega} \text{for each accessible microstate} \\ 0 \quad\quad \text{for all other micro states} \end{cases}$.

However, in any other representation, the density matrix has the following two properties:

- All the diagonal elements are equal (equal a priori probability postulate).
- All off diagonal elements are equal to zero. (A consequence of postulate of random phases).

Postulate of random phases is a requirement of the quantum mechanics. According to this ψ_k $\psi^k(t) = \sum_n a_n^k(t)\phi_n(q)$ which is a super position of the basis set $\{\phi_n\}$ is such that it is not a coherent superposition of the basis set functions but is such that $a_n^k = ae^{i\theta_n^k}$ where $e^{i\theta_m^k}$ is the random phase factor. Such that,

$$\rho_{mn}(t) = \frac{1}{\mathcal{N}}\sum_{k=1}^{\mathcal{N}} a_m^k(t)a_n^{k*}(t) = \rho_{mn}(t) = \frac{1}{\mathcal{N}}\sum_{k=1}^{\mathcal{N}} |a|^2 e^{i\left(\theta_m^k - \theta_n^k\right)}$$
$$= |a|^2 \left\langle e^{i\left(\theta_m^k - \theta_n^k\right)} \right\rangle$$

Where, $\left\langle e^{i\left(\theta_m^k - \theta_n^k\right)} \right\rangle$ is the time average equal to δ_{mn}. Thus the result is similar to equation $\rho_{mn} = \rho_n \delta_{mn}$ as expected for micro canonical ensemble.

The Canonical Ensemble

In canonical ensemble, system is in thermal equilibrium with the surroundings and is described by the three parameters N, V and T. System and surroundings can exchange energy and hence becomes a variable. The probability that the system possesses energy E_r is proportional to the Boltzmann factor $e^{-\beta E_r}$ given by:

$$P_r = \frac{e^{-\beta E_r}}{\sum_r e^{-\beta E_r}} = \frac{e^{-\beta E_r}}{Z}$$

Where Z is the partition function.

In energy representation, the density matrix elements are represented as $\rho_{mn} = \rho_n \delta_{mn}$, so that in canonical ensemble, we have,

$$\rho_{mn} = \rho_n \delta_{mn} = \frac{e^{-\beta E_n}}{Z}$$

In bracket notation the density matrix element can be written as,

$$\rho_{mn} = \langle \phi_m | \hat{\rho} | \phi_n \rangle$$

Then since,

$$\rho_{mn} = \sum_l \langle \phi_m | \phi_l \rangle \rho_l \langle \phi_l | \phi_n \rangle$$

Comparing above two equations we have,

$$\hat{\rho} = \sum_n |\phi_n\rangle \rho_n \langle \phi_n|$$

Or

$$\hat{\rho} = \sum_n |\phi_n\rangle \frac{e^{-\beta E_n}}{Z} \langle \phi_n| = \frac{e^{-\beta H}}{Z} \sum_n |\phi_n\rangle\langle\phi_n| = \frac{e^{-\beta \hat{H}}}{Z}$$
$$= \frac{e^{-\beta \hat{H}}}{\mathrm{Tr}\left(e^{-\beta \hat{H}}\right)}$$

Where, we have used the fact that $\sum_n |\phi_n\rangle\langle\phi_n|$ is the unit operator.

The expectation value of a physical quantity corresponding to the system G represented by the operator $\hat{G}$ is then given by,

$$\langle G\rangle_N = \mathrm{Tr}\left(\hat{\rho}\hat{G}\right) = \frac{\mathrm{Tr}\left(\hat{G}e^{-\beta\hat{H}}\right)}{Z} = \frac{\mathrm{Tr}\left(\hat{G}e^{-\beta\hat{H}}\right)}{\mathrm{Tr}\left(e^{-\beta\hat{H}}\right)}$$

Where following the convention, since number of particles stay constant in this ensemble sub-script explicitly describes this.

For a pure state there is only one state with $\rho_n = 1$ and using equation $\hat{\rho} = \sum_n |\phi_n\rangle\rho_n\langle\phi_n|$ density matrix is given by,

$$\hat{\rho} = |\phi\rangle\langle\phi|$$

This expression can be used for finding $\hat{\rho}$ of the pure states and density matrix of the mixed states made from a number of pure states with suitable proportion of the mixture.

The Grand Canonical Ensemble

The system can exchange both energy and particles, Keeping parameters μ, v and T constant. The Probability of the system of grand canonical ensemble to be in a state r with energy E_r with number of particles N_r is given by,

$$P_r = \frac{e^{-\beta(E_r - \mu Nr)}}{\sum_r e^{-\beta(E_r - \mu Nr)}} = \frac{e^{-\beta(E_r - \mu Nr)}}{\mathbb{Z}}$$

Where, $\mathbb{Z} = \sum_r e^{-\beta(E_r - \mu Nr)} = \mathrm{Tr}\left(e^{-\beta(\hat{H} - \mu\hat{n})}\right)$ stands for grand partition function, and $\hat{n}$ is the number operator.

Here we must note that density operator shall not only commute with Hamiltonian $\hat{H}$ but also with number operator $\hat{n}$ with eigen values $N_i = 0,1,2,\ldots\ldots\ldots\infty$.

Thus density operator can be written as,

$$\hat{\rho} = \frac{e^{-\beta(\hat{H} - \mu\hat{n})}}{\mathrm{Tr}\left(e^{-\beta(\hat{H} - \mu\hat{n})}\right)}$$

The ensemble average of a physical quantity G of the system can now be written as,

$$\langle G\rangle = \mathrm{Tr}\left(\hat{\rho}\hat{G}\right) = \frac{\mathrm{Tr}\left(\hat{G}e^{-\beta(\hat{H} - \mu\hat{n})}\right)}{\mathrm{Tr}\left(e^{-\beta(\hat{H} - \mu\hat{n})}\right)}$$

Calculation of Density Matrix: Some Prototype Applications

In the following we look at some problems where we apply the procedure developed above to calculate density matrix and mean value of a physical quantity of interest to us.

Electron in a Magnetic Field

As a simple problem, let us consider an electron in a magnetic field, $\vec{B}$ applied along z-axis. An electron has an intrinsic spin $\vec{S} = \frac{1}{2}\hbar\vec{\sigma}$, with magnetic moment $\vec{\mu} = \frac{e\hbar}{2mc}\vec{\sigma} = \mu_B\vec{\sigma}$, where $\mu_B = \frac{e\hbar}{2mc}$ is Bohr magnet on and $\vec{\sigma}$ is the Pauli spin operator with components,

$$\hat{\sigma}_x = \begin{pmatrix} 0 & 1 \\ 1 & 0 \end{pmatrix}, \hat{\sigma}_y = \begin{pmatrix} 0 & -i \\ i & 0 \end{pmatrix}, \hat{\sigma}_z = \begin{pmatrix} 1 & 0 \\ 0 & -1 \end{pmatrix}$$

In the magnetic field $\vec{B}$ the electron has only two spin orientations, one parallel to the magnetic field and the other perpendicular to the magnetic field. The configurational Hamiltonian of the spin in the magnetic field is,

$$\hat{H} = -\mu_B\vec{\sigma}.\vec{B} = -\mu_B\hat{\sigma}_Z B$$

This is equivalent to a two state problem, with energy of the two states as,

$$E_1 = -\mu_B B, E_2 = \mu_B B,$$

The density matrix for the canonical ensemble would be,

$$\hat{\rho} = \frac{e^{-\beta\hat{H}}}{\mathrm{Tr}\left(e^{-\beta\hat{H}}\right)} = \frac{e^{\beta\mu_B\hat{\sigma}_Z B}}{\mathrm{Tr}\left(e^{\beta\mu_B\hat{\sigma}_Z B}\right)}$$

Where,

$$e^{\beta\mu_B\hat{\sigma}_Z B} = \begin{pmatrix} e^{\beta\mu_B B} & 0 \\ 0 & e^{-\beta\mu_B B} \end{pmatrix}$$

Therefore,

$$\hat{\rho} = \frac{1}{e^{\beta\mu_B B} + e^{-\beta\mu_B B}} = \begin{pmatrix} e^{\beta\mu_B B} & 0 \\ 0 & e^{-\beta\mu_B B} \end{pmatrix}$$

We can use this result to find the expectation value of μ_z,

$$\left\langle \widehat{\mu_z}\,\widehat{\sigma_z} \right\rangle = \mathrm{Tr}\left(\mu_B\widehat{\sigma_z}\hat{\rho}\right) = \mu_B \mathrm{Tr}\left(\widehat{\sigma_z}\hat{\rho}\right) = \mu_B \frac{e^{\beta\mu_B B} - e^{\beta\mu_B B}}{e^{\beta\mu_B B} + e^{-\beta\mu_B B}}$$
$$= \mu_B \tanh(\beta\mu_B B)$$

Free Particle in a Box

It is an interesting problem which is taught in every elementary course of quantum mechanics, a free particle in cubical box of side L. The Hamiltonian operator of the system is,

$$\hat{H} = -\frac{\hbar^2}{2m}\nabla^2 = -\frac{\hbar^2}{2m}\left(\frac{\partial^2}{\partial x^2}+\frac{\partial^2}{\partial y^2}+\frac{\partial^2}{\partial z^2}\right)$$

With eigen-functions of the Hamiltonian operator satisfying periodic boundary conditions,

$$\phi(x+L,y,z)=\phi(x+L,y,z)=\phi(x+L,y,z)=\phi(x,y,z)$$

Which gives,

$$\phi_E(r)=\frac{1}{L^{\frac{3}{2}}}e^{i\vec{k}.\vec{r}}$$

With eigen values,

$$E=\frac{\hbar^2k^2}{2m},$$

Where k takes on values $k=(k_x,k_y,k_z)=(n_x,n_y,n_z)$, with quantum numbers n_x,n_y,n_z taking values $0,\pm1,\pm2,\pm3,\ldots\ldots$ With the knowledge of the energy wave functions, we write the density matrix $\hat{\rho}$ in canonical ensemble in co-ordinate representation as,

$$\langle\vec{r}|\hat{\rho}|\vec{r}'\rangle=\left\langle\vec{r}\left|\frac{e^{-\beta\hat{H}}}{Tr\left(e^{-\beta\hat{H}}\right)}\right|\vec{r}'\right\rangle=\frac{\langle\vec{r}|e^{-\beta\hat{H}}|\vec{r}'\rangle}{Tr\left(e^{-\beta\hat{H}}\right)}=\frac{\langle\vec{r}|e^{-\beta\hat{H}}|\vec{r}'\rangle}{\int\langle\vec{r}|e^{-\beta\hat{H}}|\vec{r}'\rangle d^3r}$$

In the light of above equation, we first calculate the numerator,

$$\langle\vec{r}|e^{-\beta\hat{H}}|\vec{r}'\rangle=\sum_E\langle\vec{r}|E\rangle e^{-\beta\hat{H}}\langle E|\vec{r}'\rangle=\sum_E e^{-\beta E}\phi_E(\vec{r})\phi_E(\vec{r}')$$

The above equations can be written as,

$$\langle\vec{r}|e^{-\beta\hat{H}}|\vec{r}'\rangle=\frac{1}{L^3}\sum_k e^{-\frac{\beta\hbar^2}{2m}k^2+ik.(\vec{r}-\vec{r}')}$$

Since $\sum_k\frac{1}{L^3}\to\frac{1}{(2\pi)^3}\int d^3k$, above equation becomes,

$$\langle\vec{r}|e^{-\beta\hat{H}}|\vec{r}'\rangle\approx\frac{1}{(2\pi)^3}\int_{-\infty}^{+\infty}e^{-\frac{\beta\hbar^2}{2m}k^2+ik.(\vec{r}-\vec{r}')}d^3k$$

The integral in the above integral can be evaluated by noting that,

$$\frac{1}{\left(2\pi^3\right)}\int e^{-\frac{\beta\hbar^2}{2m}k^2+ik.(\vec{r}-\vec{r}')}d^3k$$

$$=\frac{1}{(2\pi)^3}\left[\int_{-\infty}^{+\infty}e^{-\frac{\beta\hbar^2}{2m}k^2}\cos k.(\vec{r}-\vec{r}')d^3k+i\int_{-\infty}^{+\infty}e^{-\frac{\beta\hbar^2}{2m}k^2}\sin k.(\vec{r}-\vec{r}')d^3k\right]$$

Here one can note that first term is an even function of k and the second term is an odd function of k, therefore, since second integral vanishes and first integral can be written as $2\int_0^{+\infty}e^{-\frac{\beta\hbar^2}{2m}k^2}\cos k.(\vec{r}-\vec{r}')d^3k=\left(\frac{2m\pi}{\beta\hbar^2}\right)^{\frac{3}{2}}e^{-\frac{(|\vec{r}-\vec{r}'|)^2m}{2\beta\hbar^2}}$ and hence above equation can be written as,

$$\left\langle\vec{r}\left|e^{-\beta\hat{H}}\right|\vec{r}'\right\rangle=\left(\frac{1}{2\pi}\right)^3\int e^{-\frac{\beta\hbar^2}{2m}k^2+ik.(\vec{r}-\vec{r}')}d^3k$$

$$=\left(\frac{m}{2\pi\beta\hbar^2}\right)^{\frac{3}{2}}e^{-\frac{(|\vec{r}-\vec{r}'|)^2m}{2\beta\hbar^2}}$$

Furthermore, denominator in equation $\langle\vec{r}|\hat{\rho}|\vec{r}'\rangle=\left\langle\vec{r}\left|\frac{e^{-\beta\hat{H}}}{Tr\left(e^{-\beta\hat{H}}\right)}\right|\vec{r}'\right\rangle=\frac{\langle\vec{r}|e^{-\beta\hat{H}}|\vec{r}'\rangle}{Tr\left(e^{-\beta\hat{H}}\right)}=\frac{\langle\vec{r}|e^{-\beta\hat{H}}|\vec{r}'\rangle}{\int\langle\vec{r}|e^{-\beta\hat{H}}|\vec{r}'\rangle d^3r}$ can be written, since r = r' as,

$$\int\langle\vec{r}|e^{-\beta\hat{H}}|\vec{r}'\rangle d^3r=\left(\frac{m}{2\pi\beta\hbar^2}\right)^{\frac{3}{2}}\int d^3r=V\left(\frac{m}{2\pi\beta\hbar^2}\right)^{\frac{3}{2}}$$

Thus the matrix element of the density operator $\langle\vec{r}|\hat{p}|\vec{r}'\rangle$ using above two equations can be written as,

$$\langle\vec{r}|\hat{p}|\vec{r}'\rangle=\frac{1}{V}e^{-\frac{(|\vec{r}-\vec{r}|)^2m}{2\beta\hbar^2}}$$

The density operator matrix has some interesting properties:

- $\langle\vec{r}|\hat{p}|\vec{r}'\rangle=\langle\vec{r}'|\hat{p}|\vec{r}\rangle$, i.e. density matrix is symmetric between the states $\vec{r}$ and r'.
- The diagonal element $\langle\vec{r}|\hat{p}|\vec{r}'\rangle=\frac{1}{V}$ is independent of $\vec{r}$ implying that single particle is equally likely to be anywhere in the box.
- The coefficient of $\left(|\vec{r}-|\vec{r}'\right)^2$ in the exponential part of matrix element, $\frac{\hbar^2}{mk_BT}$, has dimensions of reciprocal of the length square and represent mean thermal wave length square and represent the spread of the wave packet and measure an uncertainty in the location of the particle in the box.

- As $\beta \to 0$ the matrix element approaches the behavior of the Dirac-delta function leading to classical behavior of the system as a point particle.

Now to get the expectation value of $\hat{H}$ we need to calculate,

$$\langle \hat{H} \rangle = \frac{Tr\left(\hat{H}e^{-\beta\hat{H}}\right)}{Tr\left(e^{-\beta\hat{H}}\right)} = -\frac{\partial\left(\ln Tr\left(e^{-\beta\hat{H}}\right)\right)}{\partial\beta}$$

Using equation $\int \langle \vec{r} | e^{-\beta\hat{H}} | \vec{r}' \rangle d^3 r = \left(\frac{m}{2\pi\beta\hbar^2}\right)^{\frac{3}{2}} \int d^3r = V\left(\frac{m}{2\pi\beta\hbar^2}\right)^{\frac{3}{2}}$ we have,

$$\langle \hat{H} \rangle = -\frac{\partial\left(\ln V\left(\frac{m}{2\pi\beta\hbar^2}\right)^{\frac{3}{2}}\right)}{\partial\beta} = \frac{3}{2}\frac{\partial(\ln\beta)}{\partial\beta} = \frac{3}{2}k_B T$$

Linear Harmonic Oscillator

As the next example let us calculate the density matrix element of the one dimensional Linear Harmonic Oscillator in the q representation. The Hamiltonian of the linear harmonic oscillator is given by,

$$\hat{H} = -\frac{\hbar^2}{2m}\frac{\partial^2}{\partial q^2} + \frac{1}{2}m\omega^2 q^2$$

With eigen values and eigen functions given by,

$$E_n = \left(n + \frac{1}{2}\hbar\omega\right), \; n = 0,1,2,\dots \dots \dots \dots$$

$$\phi_n(q) = \left(\frac{m\omega}{\pi\hbar}\right)^{\frac{1}{4}} \frac{H_n(\xi)}{\left(2^n n!\right)^{\frac{1}{2}}} e^{-\frac{1}{z}\xi^2}$$

Where $\xi = \left(\frac{m\omega}{\hbar}\right)^{\frac{1}{2}} q$ and $H_n(\xi) = (-1)^n e^{\xi^2} \frac{d^n\left(e^{\xi^2}\right)}{d\xi^n}$ is the Hermite polynomial. The nth derivative of e^{ξ^2} can be easily evaluated by using the Fourier transform relation of e^{ξ^2}.

$$e^{\xi^2} = \frac{1}{\pi^{\frac{1}{2}}}\int_{-\infty}^{\infty} e^{\left(-u^2+2i\xi u\right)} du$$

So Hermite polynomial in integral form can be written as,

$$H_n(\xi) = (-1)^n e^{\xi^2} \frac{(2i)^n}{\pi^{\frac{1}{2}}}\int_{-\infty}^{\infty} u^n e^{\left(-u^2+2i\xi u\right)} du$$

So the density matrix in q representation can be written as,

$$\langle q|\hat{\rho}|q'\rangle = \sum_{n=0}^{\infty} e^{-\beta E_n}\phi_n(q)\phi_n(q')$$

$$= \left(\frac{m\omega}{\pi\hbar}\right)^{\frac{1}{2}} e^{-\frac{1}{2}(\xi^2+\xi'^2)} \sum_{n=0}^{\infty} e^{-\beta E_n} \frac{H_n(\xi)H_n(\xi')}{2^n n!}$$

$$= \left(\frac{m\omega}{\pi\hbar}\right)^{\frac{1}{2}} e^{-\frac{\beta\hbar\omega}{2}} e^{-\frac{1}{2}(\xi^2+\xi'^2)} \sum_{n=0}^{\infty} e^{-\beta(n\hbar\omega)} \frac{H_n(\xi)H_n(\xi')}{2^n n!}$$

Using equation $H_n(\xi) = (-1)^n e^{\xi^2} \frac{(2i)^n}{\pi^{\frac{1}{2}}} \int_{-\infty}^{\infty} u^n e^{(-u^2+2i\xi u)} du$ and above equation can be expressed as,

$$\langle q|\hat{\rho}|q'\rangle = \left(\frac{m\omega}{\pi^3\hbar}\right)^{\frac{1}{2}} e^{-\frac{\beta\hbar\omega}{2}} e^{\frac{1}{2}(\xi^2+\xi'^2)}$$

$$\int_{-\infty}^{\infty}\int_{-\infty}^{\infty}\sum_{n=0}^{\infty} \frac{(-2)^n}{n!} \left(e^{-\beta\hbar\omega}uv\right)^n e^{(-u^2-v^2+2i(\xi u+\xi' v))} dudv$$

In above equation integral involving summation can be solved, let us look at the summation first,

$$\sum_{n=0}^{\infty} \frac{(2)^n}{n!} \left(e^{-\beta\hbar\omega}uv\right)^n = \sum_{n=0}^{\infty} \frac{\left(-2e^{-\beta\hbar\omega}uv\right)^n}{n!} = e^{(-e^{-\beta\hbar\omega}2uv)}$$

Using above equation and $\int_{-\infty}^{\infty}\int_{-\infty}^{\infty}\sum_{n=0}^{\infty} \frac{(-2)^n}{n!} \left(e^{-\beta\hbar\omega}uv\right)^n e^{(-u^2-v^2+2i(\xi u+\xi' v))} dudv$ can be written as,

$$\langle q|\hat{\rho}|q'\rangle = \left(\frac{m\omega}{\pi^3\hbar}\right)^{\frac{1}{2}} e^{-\frac{\beta\hbar\omega}{2}} e^{\frac{1}{2}(\xi^2+\xi'^2)}$$

$$\int_{-\infty}^{\infty}\int_{-\infty}^{\infty} e^{(-u^2-v^2+2i(\xi u+\xi' v)-2uve^{-\beta\hbar\omega})} dudv$$

We state general integral formula,

$$\int \ldots \int_{-\infty}^{\infty} e^{\left[-\frac{1}{2}\sum_j^n \sum_k^n a_{jk}x_j x_k + i\sum_k^n b_k x_k\right]} dx_1 \ldots\ldots\ldots dx_n$$

$$= \frac{(2\pi)^{\frac{n}{2}}}{\left[\det(A)\right]^{\frac{1}{2}}} e^{\left(-\frac{1}{2}\sum A_{jk}^{-1} b_j b_k\right)}$$

Comparing integrand in equation $\langle q|\hat{\rho}|q'\rangle = \left(\frac{m\omega}{\pi^3\hbar}\right)^{\frac{1}{2}} e^{-\frac{\beta\hbar\omega}{2}} e^{\frac{1}{2}(\xi^2+\xi'^2)} \int_{-\infty}^{\infty}\int_{-\infty}^{\infty} e^{(-u^2-v^2+2i(\xi u+\xi' v)-2uve^{-\beta\hbar\omega})} dudv$

with integrand in above equation we have k = 1, 2, j = 1, 2 $x_1 = u$, $x_2 = v$, n = 2; $a_{11} = 1$ $a_{12} = e^{-\beta\hbar\omega}$ $a_{21} = e^{-\beta\hbar\omega}$ $a_{22} = 1$; $b_1 = 2\xi$ $b_2 = 2\xi'$, $\det A = 4\left(1-e^{-2\beta\hbar\omega}\right)$ and $A^{-1} = \frac{1}{2\left(1-e^{-2\beta\hbar\omega}\right)}\begin{bmatrix} 1 & -e^{-\beta\hbar\omega} \\ -e^{-\beta\hbar\omega} & 1 \end{bmatrix}$

Therefore,

$$\int_{-\infty}^{\infty}\int_{-\infty}^{\infty} e\left(-u^2 - v^2 + 2i(\xi u + \xi' v) - 2uve^{-\beta\hbar\omega}\right)dudv$$

$$= \frac{\pi}{\left(1-e^{-2\beta\hbar\omega}\right)^{\frac{1}{2}}}\left[e^{-\left(\frac{1}{\left(1-e^{-2\beta\hbar\omega}\right)}\right)\left(\xi^2 - 2e^{-\beta\hbar\omega}\xi\xi' + \xi'^2\right)}\right]$$

$$\langle q|\hat{\rho}|q'\rangle = \left(\frac{m\omega}{\pi\hbar}\right)^{\frac{1}{2}}$$

$$\frac{e^{-\frac{\beta\hbar\omega}{2}}}{\left(1-e^{-2\beta\hbar\omega}\right)^{\frac{1}{2}}}\left[e^{\left[\frac{1}{2}\left(\xi^2+\xi'^2\right) - \left(\frac{1}{\left(1-e^{-2\beta\hbar\omega}\right)}\right)\left(\xi^2 - 2e^{-\beta\hbar\omega}\xi\xi' + \xi'^2\right)\right]}\right]$$

Or

$$\langle q|\hat{\rho}|q'\rangle = \left(\frac{m\omega}{2\pi\hbar\sinh(\beta\hbar\omega)}\right)^{\frac{1}{2}}\left[e^{\left[-\frac{1}{2}\left(\xi^2+\xi'^2\right)\coth(\beta\hbar\omega) - \left(\frac{\xi\xi'}{\sinh(\beta\hbar\omega)}\right)\right]}\right]$$

Putting the value of ξ and ξ' above equation becomes,

$$\langle q|\hat{\rho}|q'\rangle$$

$$= \left(\frac{m\omega}{2\pi\hbar\sinh(\beta\hbar\omega)}\right)^{\frac{1}{2}}\left[e^{\left(-\frac{m\omega}{4\hbar}\right)\left[2\left(q^2+q'^2\right)\coth(\beta\hbar\omega) + \left(\frac{4qq'}{\sinh(\beta\hbar\omega)}\right)\right]}\right]$$

Recall $2\left(q^2+q'^2\right) = (q+q')^2 + (q-q')^2$ and $4qq' = (q+q')^2 - (q-q')^2$

$$2\left(q^2+q'^2\right)\coth(\beta\hbar\omega) + \left(\frac{4qq'}{\sinh(\beta\hbar\omega)}\right)$$

$$= (q+q')^2\left(\coth\beta\hbar\omega + \frac{1}{\sinh\beta\hbar\omega}\right) + (q-q')^2\left(\coth\beta\hbar\omega - \frac{1}{\sinh\beta\hbar\omega}\right)$$

$$= (q+q')^2\left(\tanh\frac{\beta\hbar\omega}{2}\right) + (q-q')^2\left(\coth\frac{\beta\hbar\omega}{2}\right)$$

Density matrix element in equation $\langle q|\hat{\rho}|q'\rangle = \left(\frac{m\omega}{2\pi\hbar\sinh(\beta\hbar\omega)}\right)^{\frac{1}{2}}\left[e^{\left(-\frac{m\omega}{4\hbar}\right)\left[2\left(q^2+q'^2\right)\coth(\beta\hbar\omega)+\left(\frac{4qq'}{\sinh(\beta\hbar\omega)}\right)\right]}\right]$ can be written as,

$$\langle q|\hat{\rho}|q'\rangle$$

$$= \left(\frac{m\omega}{2\pi\hbar\sinh(\beta\hbar\omega)}\right)^{\frac{1}{2}}\left[e^{\left(-\frac{m\omega}{4h}\right)\left[(q+q')^2\left(\tanh\frac{\beta\hbar\omega}{2}\right)+(q-q')^2\left(\coth\frac{\beta\hbar\omega}{2}\right)\right]}\right]$$

Now we can calculate the partition function $\mathrm{Tr}\left(e-\beta\hat{H}\right)$:

$$\mathrm{Tr}\left(e-\beta\hat{H}\right) = \int\langle q|\hat{\rho}|q\rangle dq$$

$$= \left(\frac{m\omega}{2\pi\hbar\sin h(\beta\hbar\omega)}\right)^{\frac{1}{2}}\int\left[e^{\left(-\frac{4m\omega}{4\hbar}\right)\left[(q)^2\left(\tanh\frac{\beta\hbar\omega}{2}\right)\right]}\right]dq$$

$$= \frac{1}{2\sin h\left(\frac{\beta\hbar\omega}{2}\right)} = \frac{e-\frac{\beta\hbar\omega}{2}}{1-e^{-\beta\hbar\omega}}$$

Interesting once again we have arrived at the partition function of a linear harmonic oscillator.

We can calculate the mean energy of the harmonic oscillator $\langle\hat{H}\rangle$

$$\langle\hat{H}\rangle = -\frac{\partial}{\partial\beta}\ln\mathrm{Tr}\left(e^{-\beta\hat{H}}\right) = \frac{\hbar\omega}{2}\coth\frac{\beta\hbar\omega}{2}$$

Equation $\langle q|\hat{\rho}|q'\rangle = \left(\frac{m\omega}{2\pi\hbar\sinh(\beta\hbar\omega)}\right)^{\frac{1}{2}}\left[e^{\left(-\frac{m\omega}{4\hbar}\right)\left[(q+q')^2\left(\tanh\frac{\beta\hbar\omega}{2}\right)+(q-q')^2\left(\coth\frac{\beta\hbar\omega}{2}\right)\right]}\right]$ provide us the probability density for the linear harmonic oscillator coordinate near q is,

$$\langle q|\hat{\rho}|q\rangle = \left(\frac{m\omega}{2\pi\hbar\sinh(\beta\hbar\omega)}\right)^{\frac{1}{2}}\left[e^{\left(-\frac{m\omega q^2}{h}\right)\left(\tanh\frac{\beta\hbar\omega}{2}\right)}\right]$$

It is a Gaussian distribution in q, with mean value zero and root mean square deviation,

$$q_{r.m.s} = \left[\frac{\hbar}{2m\omega\tanh\frac{\beta\hbar\omega}{2}}\right]^{\frac{1}{2}}$$

For the classical limiting case $\beta\hbar\omega<<1$, $\sinh(\beta\hbar\omega)\rightarrow\beta\hbar\omega$ and $\tanh\frac{\beta\hbar\omega}{2}\rightarrow\beta\hbar\omega$, probability density becomes,

$$\langle q|\hat{\rho}|q\rangle = \left(\frac{m\omega^2 k_B T}{2\pi\hbar^2\omega^2}\right)^{\frac{1}{2}} \left[e^{\left(-\frac{m\omega^2 q^2}{k_B T}\right)} \right]$$

The density matrix calculated above also allows us to calculate $\langle q^2 \rangle$ and $\langle p^2 \rangle$.

Recalling that,

$$\frac{\int_{-\infty}^{\infty} x^2 e^{-\frac{1}{2}\alpha x^2} dx}{\int_{-\infty}^{\infty} e^{-\frac{1}{2}\alpha x^2}} = \frac{d}{d\alpha} \ln \int_{-\infty}^{\infty} e^{-\frac{1}{2}\alpha x^2} = \frac{1}{\alpha}$$

$$\langle q^2 \rangle = \frac{\int_{-\infty}^{\infty} q^2 \langle q|\hat{\rho}|q\rangle dq}{\int_{-\infty}^{\infty} \langle q|\hat{\rho}|q\rangle dq} = \frac{d}{d\alpha} \ln \left[C \int_{-\infty}^{\infty} e^{-\frac{1}{2}\alpha q^2} \right]$$

$$= \frac{1}{m\omega^2} \frac{\hbar\omega}{2} \coth\left(\frac{1}{2}\beta\hbar\omega\right) = \frac{1}{m\omega^2} \langle \hat{H} \rangle$$

Where $\alpha = \left(\frac{2m\omega}{\hbar}\right)\left(\tanh\frac{\beta\hbar\omega}{2}\right)$ and $C = \left(\frac{m\omega}{2\pi\hbar\sinh(\beta\hbar\omega)}\right)^{\frac{1}{2}}$

Or

$$\frac{1}{2} m\omega^2 \langle q^2 \rangle = \frac{1}{2} \langle \hat{H} \rangle$$

Thus potential energy of the oscillator is $\frac{1}{2}\langle \hat{H} \rangle$

Similarly one can show that,

$$\langle p^2 \rangle = \left\langle -\frac{\hbar^2}{2m} \frac{\partial^2}{\partial q^2} \right\rangle = m \langle \hat{H} \rangle$$

Or $\frac{\langle p^2 \rangle}{2m} = \frac{1}{2}\langle \hat{H} \rangle$ i.e. Mean value of kinetic energy and potential energy of the harmonic oscillator are the same and equal to half the total mean energy.

Indistinguishability of Particles and Quantum Statistical Mechanics

Till this point to take care of indistinguishability of particles, we made use of ad hoc explanations to correct the counting of microstates as was done in the case of resolving of Gibbs Paradox earlier. However, quantum mechanical description requires us to deal with great care in describing systems composed of indistinguishable particles. The key idea lies in symmetrization of wave functions which is closely related with indistinguishability of particles leading to special correlations which are attractive in the case of particles obeying Bose-Einstein Statistics and repulsive in the case of particles obeying Fermi Dirac-Statistics. We shall study these aspects by considering a gas of N non-interacting identical particles. The Hamiltonian, $\hat{H}$ for such a system is a sum of the individual single particle Hamiltonians $\hat{H}_\iota$.

$$\hat{H}(\vec{q},\vec{p})=\sum_{i=o}^{N}\hat{H}_\iota(q_i,p_i)$$

Identical nature of particles here means that except for the values of the arguments q_i and pi, H_i are formally identical.

The time independent Schrodinger equation of the system is,

$$\hat{H}\psi_E(\vec{q})=E\psi_E(\vec{q})$$

Where $\psi_E(q)$ is a product of single particle eigenfunctions $u_{\varepsilon_i}(q_i)$ of the single particle Hamiltonian $\hat{H}_\iota(q_i,p_i)$ with energy ε_i.

$$\psi_E(\vec{q})=\prod_{i=1}^{N}u_{\varepsilon_i}(q_i)$$

With

$$E=\sum_{i=1}^{N}\varepsilon_i$$

This situation can be alternately described in terms of a distribution set $\{n_i\}$ such that n_i number of particles are in an eigenstate with energy ε_i such that,

$$\sum_i n_i = N$$

And

$$E=\sum_i n_i\varepsilon_i$$

The eigen state described by above two equations can be written as,

$$\psi_E(\vec{q})=\prod_{m=1}^{n_1}u_1(m)\prod_{m=n_1+1}^{n_1+n_2}u_2(m)\dots\dots\dots$$

Where $u_i(m)-u_{\varepsilon_i}(q_m)$.

Symmetrization of State Functions and Gibbs Paradox

So far this simple looking multi particle wave function describes the state of the system. However, we need to ask the question what happens when we permute the co-ordinates (1, 2, 3, 4 ..., N) in above equation such that new co-ordinates become (P1, P2, P3, P4 ..., N) say the resulting wave function $P\psi_E(q)$ becomes,

$$P\psi_E(\vec{q})=\prod_{m=1}^{n_1}u_1(Pm)\prod_{m=n_1+1}^{n_1+n_2}u_2(Pm)\dots\dots\dots$$

Now if the particles are distinguishable, this results in a new microstate of the system. So in the case of N distinguishable particles with n_1 particles in a state with energy ε_1 and n_2 particles in a state with energy ε_2 and so on, the for a distribution set $\{n_i\}$ the number of distinct microstates is,

$$\frac{N!}{n_1!n_2!n_3!.....}$$

This weight factor on applying Gibbs correction becomes,

$$W_C\{n_i\}\frac{1}{n_1!n_2!n_3!.....}$$

This ad hoc removal of N! was realization of the intrinsic indistinguishability of particles. Though applying of this correction led to extensive nature of entropy, quantum, mechanically permutations among identical particles should not get us a new microstate of the system. The effect of permutation therefore should not lead to any new microstate implying that the weight factor described by above equation should be identically equal to unity for a given distribution set $\{n_i\}$, provided there is no physical reason disallowing the set itself.

Therefore,

$$W_q\{n_i\}=1$$

What are the implications of this for the wave function described by equation $P\psi_E(\vec{q})=\prod_{m=1}^{n_1}u_1(Pm)\prod_{m=n_1+1}^{n_1+n_2}u_2(Pm).........$ since permutation of the arguments shall lead to a mathematically and physically distinct state. But this is not desirable since mere interchange of the co-ordinates should not lead to a new microstate.

Construction of Symmetric and Antisymmetric Wavefunctions

To achieve this, it is required that N! possible wavefunctions obtained after permutations should not lead to a new microstate thereby making $\psi_E(q)$ unaffected by these permutations. In quantum mechanics, we can do this by expressing $\psi_E(q)$ as a linear combination of all the N! wave functions such that the original wavefunction ψ and the permuted wavefunction P ψ should be such that their probability amplitude must be the same as stated below,

$$|P\psi|^2=|\psi|^2$$

This means either,

$$P\psi=\psi \text{ for all P}$$

Or

$$P\psi=\begin{cases}+\psi \text{ if P is an even permutation}\\-\psi \text{ if is an odd permutation}\end{cases}$$

Equation $P\psi = \psi$ for all P implies that wave function is symmetric and above equation implies that wave function is antisymmetric in its arguments. Mathematically, the two types of wave functions can be written as:

$$\psi_S(\vec{q}) = \text{const.}\sum_P P\psi(\vec{q})$$

And

$$\psi_A(\vec{q}) = \text{const.}\sum_P \delta_P P\psi(\vec{q})$$

Where δ_p is +1 for P to be even and is -1 for P to be odd. For example for a three particle system, symmetric wave function can be written as,

$$\psi_S(\vec{q}) = \text{const.}\left[\psi(1,2,3)+\psi(2,3,1)+\psi(3,2,1)+\psi(1,3,2)+\psi(3,2,1)+\psi(2,1,3)\right]$$

Here first three terms involve even permutations and last three odd permutations.

Similarly for a three particle system antisymmetric wave function can be written as,

$$\psi_A(\vec{q}) = \text{const.}\left[\psi(1,2,3)+\psi(2,3,1)+\psi(3,2,1)+\psi(1,3,2)-\psi(3,2,1)-\psi(2,1,3)\right]$$

In terms of single particle states $\psi(1,2,3) = u_1(1)u_2(2)u_3(3)$, $\psi(2,3,1) = u_1(2)u_2(3)u_3(1)$, ……, $\psi_A(\vec{q})$ can be expressed as a slater determinant,

$$\psi_A(\vec{q}) = \text{Const.}\begin{vmatrix} u_1(1) & u_1(2) & u_1(3) \\ u_2(1) & u_2(2) & u_2(3) \\ u_3(1) & u_3(2) & u_2(3) \end{vmatrix}$$

As one expands the determinant, the positive and negative signs appear automatically, on interchanging the arguments the antisymmetric wave function changes the sign as required. Also if two or more particle happens to be in the same state corresponding rows become identical and determinant vanishes implying that such state is not physically possible, equivalent to Pauli's exclusion principle.

In other words, we can say that a system of particles which follows exclusion principle must have an antisymmetric wave function. Such particles are said to obey Fermi Dirac statistics and are described as fermions. In this case quantum statistical weight factor described by equation $W_q\{n_i\} = 1$ now called $W_{F.D.}\{n_i\}$ is such that the as long as distribution set elements n_i have value either 0 or 1 it is unity else zero i.e.

$$W_{F.D.}\{n_i\} = \begin{cases} 1 \text{ if } \sum_i n_i^2 = N \\ 0 \text{ if } \sum_i n_i^2 > N \end{cases}$$

For systems characterized by symmetric wavefunctions, n_i can be 0 or positive integers. Such systems are said to follow Bose-Einstein Statistics and are described as bosons. The weight factor $W_{B.E.}\{n_i\}$ for such a system is identically equal to 1, whatever the value of number $\{n_i\}$:

$$W_{B.E.}\{n_i\} = 1, n_i = 0, 1, 2, \ldots \ldots$$

Spin and Statistics

The spin of a particle and statistics it obeys are intimately connected. According to relativistic quantum field theory, particles with half integer spin have got antisymmetric wave functions and particles with integer spin has wave functions which are symmetric. Accordingly particles with half integer spin follow Fermi-Dirac Statistics and particles which have integer spin follow Bose-Einstein Statistics. Examples of fermions are electrons, positrons, neutrons, protons, μ mesons. Examples of bosons are photons, π mesons. A composite particle with consituents as fermions and bosons can either be a boson or a fermion, depending on the spin of the composite particle which is resultant of the spin of the constituents. Accordingly He^3 (two electrons, two protons and one neutron) is a fermion whereas He^4 (two electrons, two protons and two neutrons) is a boson.

References

- Canonical-ensemble: uni-frankfurt.de, Retrieved 26, June 2020
- Ergodic-hypothesis-in-classical-statistical-mechanics-228878401: researchgate.net, Retrieved 04, April 2020

4

Fermi Gas: An Integrated Study

The state of matter which comprises an ensemble of various non-interacting fermions is termed as a Fermi gas. A few examples of such particles are electrons, protons and neutrons. These particles have a half-integer spin. This chapter has been written to provide a thorough understanding of the different facets of Fermi gas.

Fermi gas contains fermions. The fermions do not interact with each other but according to Pauli Exclusion Principle they interact with each other. The Fermi energy is the energy of the most energetic fermions. Basically at the absolute zero temperature, the energy of fermions is known as Fermi energy. The Fermi energy surface in momentum space is known as Fermi surface. An ideal Fermi gas is a collection of fermions. Analysis of Fermi gas nature is a quantum mechanical approach. The most popular example of Fermi gas is electrons. The Fermi gas obeys Fermi Dirac statistics. The energy distribution of fermions in a Fermi gas can be determined by the Fermi Dirac statistics in thermal equilibrium. By using the Fermi Dirac statistics, we can characterize by their number density, temperature and the set of energy level. At absolute zero temperature, the total energy of a Fermi gas is greater than the sum of total energy of a single particle at ground state. This energy difference happens due to the sort of interaction or pressure for which there is a separation is created between fermions. This sort of pressure or interaction occurs due to the Pauli Exclusion Principle. Thus, at zero temperature, the Fermi gas is having a non zero value. This Fermi pressure is known as degeneracy pressure. When a Fermi gas is consisting neutrons, is known as neutron stars. Again, when the Fermi gas is containing the electrons, is known as white dwarf. This neutron gas or white dwarf can be stabilized by the degeneracy pressure.

"Fermi level" is the term used to describe the top of the collection of electron energy levels at absolute zero temperature. This concept comes from Fermi-Dirac statistics. Electrons are fermions and by the Pauli exclusion principle cannot exist in identical energy states. So at absolute zero they pack into the lowest available energy states and build up a "Fermi sea" of electron energy states. The Fermi level is the surface of that sea at absolute zero where no electrons will have enough energy to rise above the surface. The concept of the Fermi energy is a crucially important concept for the understanding of the electrical and thermal properties of solids. Both ordinary electrical and thermal processes involve energies of a small fraction of an electron volt. But the Fermi energies of metals are on the order of electron volts. This implies that the vast majority of the electrons cannot receive energy from those processes because there are no available energy states for them to go to within a fraction of an electron volt of their present energy. Limited to a tiny depth of energy, these interactions are limited to "ripples on the Fermi sea".

At higher temperatures a certain fraction, characterized by the Fermi function, will exist above the Fermi level. The Fermi level plays an important role in the band theory of solids. In doped semi-conductors, p-type and n-type, the Fermi level is shifted by the impurities, illustrated by their band gaps. The Fermi level is referred to as the electron chemical potential in other contexts.

In metals, the Fermi energy gives us information about the velocities of the electrons which participate in ordinary electrical conduction. The amount of energy which can be given to an electron in such conduction processes is on the order of micro-electron volts, so only those electrons very close to the Fermi energy can participate. The Fermi velocity of these conduction electrons can be calculated from the Fermi energy.

$$v_F = \sqrt{\frac{2E_F}{m}}$$

This speed is a part of the microscopic Ohm's Law for electrical conduction. For a metal, the density of conduction electrons can be implied from the Fermi energy.

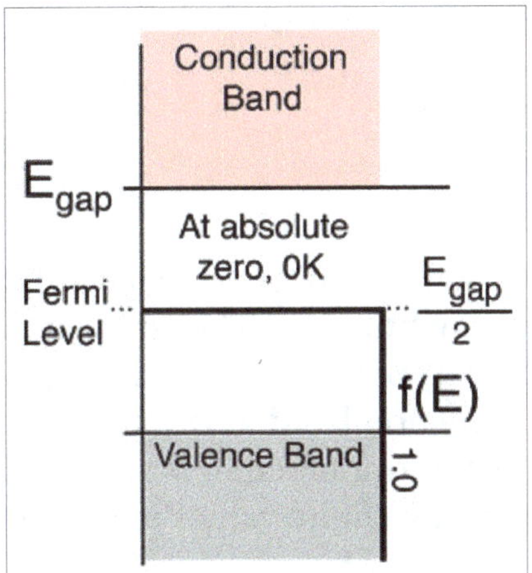

Context of Fermi level for a semiconductor

Temperatures, while they are dominant contributors to thermal conductivity and electrical conductivity. Since only a tiny fraction of the electrons in a metal are within the thermal energy kT of the Fermi energy, they are "frozen out" of the heat capacity by the Pauli principle. At very low temperatures, the electron specific heat becomes significant.

The Fermi function $f(E)$ gives the probability that a given available electron energy state will be occupied at a given temperature. The Fermi function comes from Fermi-Dirac statistics and has the form

$$f(E) = \frac{1}{e^{(E-E_F)/kT} + 1}$$

The basic nature of this function dictates that at ordinary temperatures, most of the levels up to the Fermi level EF are filled, and relatively few electrons have energies above the Fermi level. The Fermi level is on the order of electron volts (e.g., 7 eV for copper), whereas the thermal energy kT is only about 0.026 eV at 300K. If you put those numbers into the Fermi function at ordinary temperatures, you find that its value is essentially 1 up to the Fermi level, and rapidly approaches zero above it.

The illustration below shows the implications of the Fermi function for the electrical conductivity of a semiconductor. The band theory of solids gives the picture that there is a sizable gap between the Fermi level and the conduction band of the semiconductor. At higher temperatures, a larger fraction of the electrons can bridge this gap and participate in electrical conduction.

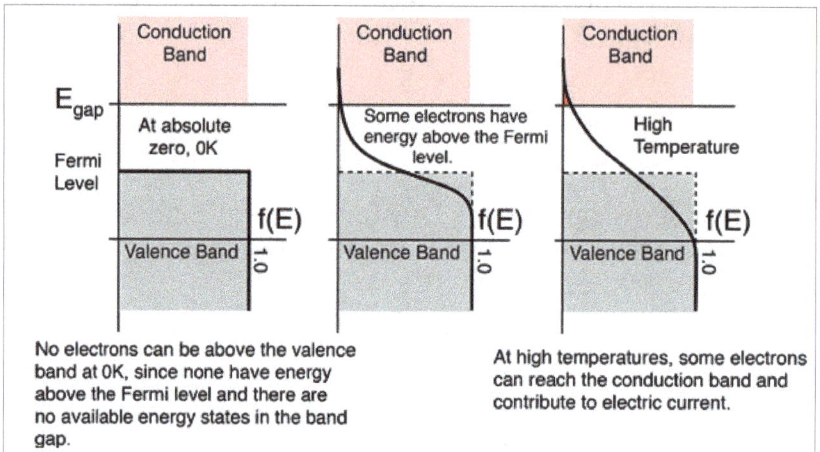

Note that although the Fermi function has a finite value in the gap, there is no electron population at those energies (that's what you mean by a gap). The population depends upon the product of the Fermi function and the electron density of states. So in the gap there are no electrons because the density of states is zero. In the conduction band at 0K, there are no electrons even though there are plenty of available states, but the Fermi function is zero. At high temperatures, both the density of states and the Fermi function have finite values in the conduction band, so there is a finite conducting population.

Free Electrons in One Dimension

Energy Levels in One Dimension

Consider a free electron gas in one dimension. An electron of mass m is confined to a length L by infinite barriers. We consider first a free electron gas in one dimension. The wave function $\psi_n(x)$ of the electron is a solution of the Schrödinger equation $H\psi_n(x) = E_n\psi_n(x)$, where E_n is the energy of electron orbital. Since we can assume that the potential lies at zero, the Hamiltonian H includes only the kinetic energy so that

$$H\psi_n(x) = -\frac{\hbar^2}{2m}\frac{d^2}{dx^2}\psi_n(x) = E_n\psi_n(x)$$

Note that this is a one-electron equation, which means that we neglect the electron-electron interactions. We use the term orbital to describe the solution of this equation. Since the $\Psi_n(x)$ is a continuous function and is equal to zero beyond the length L, the boundary conditions for the wave function are $\psi_n(0)=\psi_n(L)$. The solution of equation is therefore,

$$\psi_n(x)=A\sin\left(\frac{n\pi}{L}x\right)$$

where A is a constant and n is an integer. Substituting above equations we obtain for the eigenvalues,

$$E_n(x)=\frac{\hbar^2}{2m}\left(\frac{n\pi}{L}\right)^2.$$

These solutions correspond to standing waves with a different number of nodes within the potential well as is shown in figure.

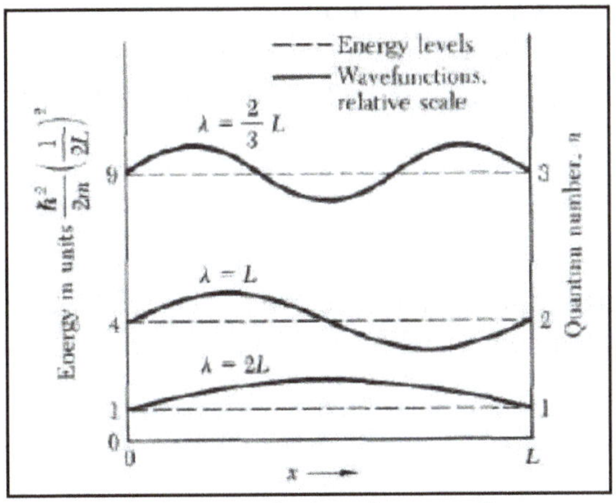

Figure: First three energy levels and wave-functions of a free electron of mass m confined to a line of length L. The energy levels are labeled according to the quantum number n which gives the number of half-wavelengths in the wave function. The wavelengths are indicated on the wave functions.

Dispersion Relations of One Dimension

The energy of the confined electron is related to its wave vector k.

$$E_n(k)=\frac{\hbar^2k^2}{2m}$$

So the dispersion relation of the confined electron is displayed below.

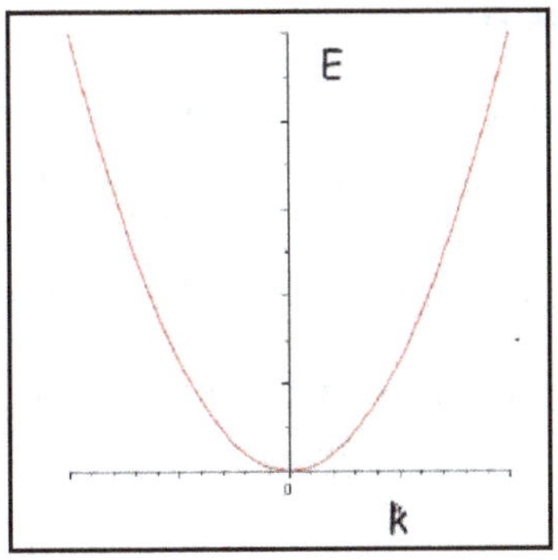

Figure: Dispersion relation for one-dimensional electron system.

Fermi Energy in One Dimension

Now we need to accommodate N valence electrons in these quantum states. According to the Pauli exclusion principle no two electrons can have their quantum number identical. That is, each electronic quantum state can be occupied by at most one electron. The electronic state in a 1D solid is characterized by two quantum numbers that are n and m_s, where n describes the orbital $\psi_n(x)$, and ms describes the projection of the spin momentum on a quantization axis. Electron spin is equal to $S=1/2$, so that there $(2S+1)=2$ possible spin states with $m_s = \pm½$. Therefore, each orbital labeled by the quantum number n can accommodate two electrons, one with spin up and one with spin down orientation.

Let n_F denote the highest filled energy level, where we start filling the levels from the bottom $(n=1)$ and continue filling higher levels with electrons until all N electrons are accommodated. It is convenient to suppose that N is an even number. The condition $2n_F = N$ determines n_F, the value of n for the uppermost filled level. The energy of the highest occupied level is called the Fermi energy E_F. For the one dimensional system of N electrons we find, using equation,

$$E_F = \frac{\hbar^2}{2m}\left(\frac{N\pi}{2L}\right)^2.$$

In metals the value of the Fermi energy is of the order of 5 eV. The ground state of the N electron system is illustrated: All the electronic levels are filled up to the Fermi energy. All the levels above are empty.

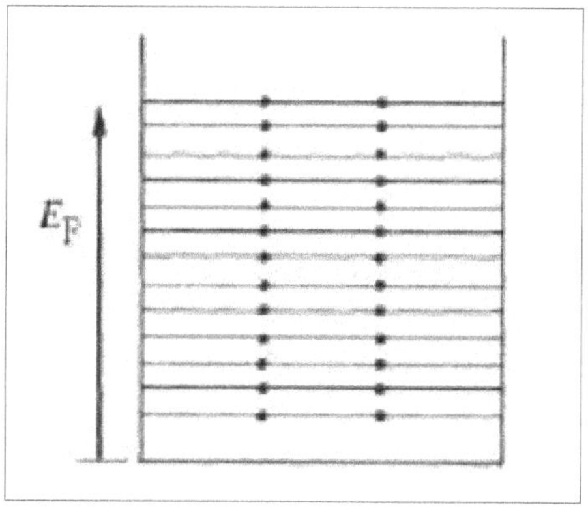

Figure: Occupation of energy levels according to the Pauli exclusion principle.

Density of states in One Dimension

To calculate various optical properties such as the rate of absorption or emission and how electrons and holes distribute themselves within a solid, we need to know the number of available states per unit volume per unit energy. We first calculate the available states in k-space and then use the energy-momentum relation in parabolic bands to give the density of states in terms of energy.

In one dimensional case, we have derived its discrete wave vector k_n

$$k_n = \frac{\pi n}{L}$$

This equation suggests that there is one quantum state within the momentum interval π / L. Consequently the number of quantum states N in the momentum interval dk is

$$dN = \frac{dk}{\pi / L} = \frac{L}{\pi} dk$$

Multiplying by 2 for spin degeneracy, the number of states within the interval dk in 1-d,

$$dN = 2 \times \frac{dk}{\pi / L} = \frac{2L}{\pi} dk$$

For one dimensional quantum wires, we have

$$E = \frac{\hbar^2 k^2}{2m}$$

$$dk = \sqrt{\frac{m}{2\hbar^2 E}} dE$$

Therefore the density of states per unit length in 1-d and multiplying by 2 for spin degeneracy,

$$dN = \frac{2L}{\pi} dk = \frac{2L}{\pi} \sqrt{\frac{m}{2\hbar^2 E}} dE$$

$$DOS = \frac{dN}{dE} = \frac{L}{\pi} \sqrt{\frac{2m}{\hbar^2}} \frac{1}{\sqrt{E}}$$

For one dimension, the density of states per unit volume at energy E is given by

$$DOS = \frac{dN}{dE} = \frac{L}{\pi} \sqrt{\frac{2m}{\hbar^2}} \frac{1}{\sqrt{E}}$$

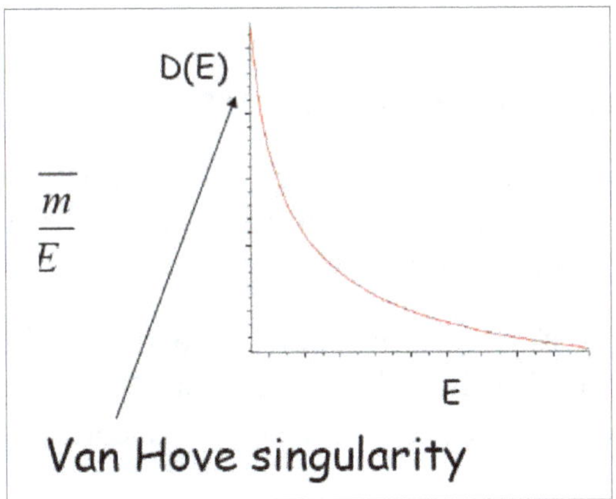

Figure: DOS of one dimensional system.

Using more than the first energy level, the density of states function becomes

$$DOS = \frac{L}{\pi} \sqrt{\frac{2m}{\hbar^2}} \sum_i \frac{n_i H(E - E_i)}{\sqrt{E - E_i}}$$

where once again, $H(E-E_i)$ is the Heaviside function and n_i is the degeneracy factor. For quantum structures with dimensions lower than 2, it is possible for the same energy level to occur for more than one arrangement of confined states. To account for this, a second factor $n_i(E)$ is introduced.

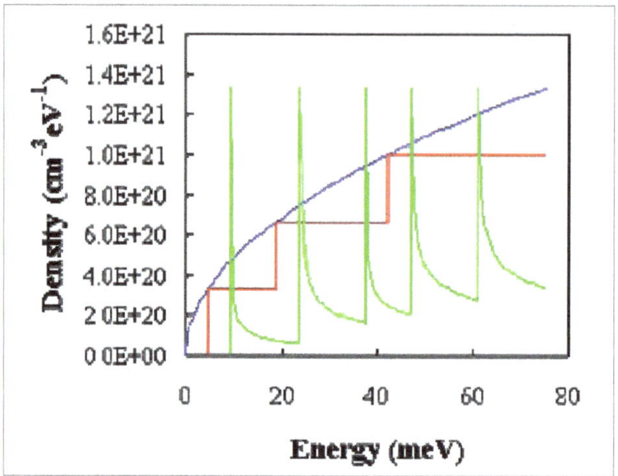

Figure: Density of states for quantum wire (1D green) which has multiple states.

Fermi Dirac Distribution

Distribution functions are nothing but the probability density functions used to describe the probability with which a particular particle can occupy a particular energy level. When we speak of Fermi-Dirac distribution function, we are particularly interested in knowing the chance by which we can find a fermion in a particular energy state of an atom. Here, by fermions, we mean the electrons of an atom which are the particles with ½ spin, bound to Pauli exclusion principle.

Necessity of Fermi Dirac Distribution Function

In fields like electronics, one particular factor which is of prime importance is the conductivity of materials. This characteristic of the material is brought about the number of electrons which are free within the material to conduct electricity.

As per energy band theory, these are the number of electrons which constitute the conduction band of the material considered. Thus in order to have an idea over the conduction mechanism, it is necessary to know the concentration of the carriers in the conduction band.

Fermi Dirac Distribution Expression

Mathematically the probability of finding an electron in the energy state E at the temperature T is expressed as

$$f(E) = \frac{1}{1 + e^{\left(\frac{E - E_f}{K_B T}\right)}}$$

Where,

$$K_B = 1.38 \times 10^{-23}\ J\,K^{-1}$$

is the Boltzmann constant

T is the absolute temperature

E_f is the Fermi level or the Fermi energy

Effect of Temperature on Fermi-Dirac Distribution Function

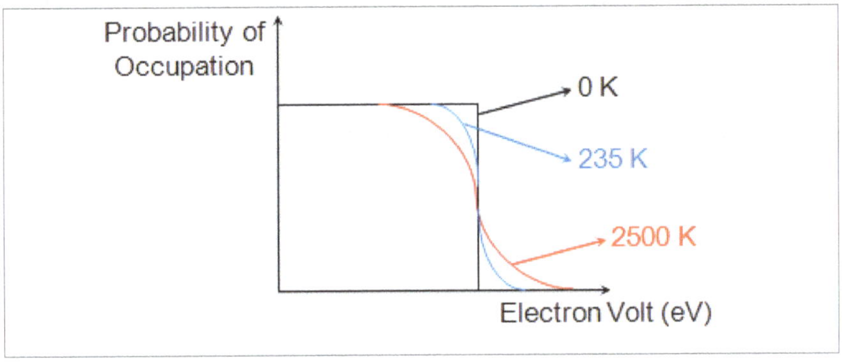

Figure: Fermi-Dirac Distribution Function at Different Temperature

At $T = 0\,K$, the electrons will have low energy and thus occupy lower energy states. The highest energy state among these occupied states is referred to as Fermi-level. This in turn means that no energy states which lie above the Fermi-level are occupied by electrons. Thus we have a step function defining the Fermi-Dirac distribution function as shown by the black curve in figure above. However as the temperature increases, the electrons gain more and more energy due to which they can even rise to the conduction band. Thus at higher temperatures, one cannot clearly distinguish between the occupied and the unoccupied states as indicated by the blue and the red curves shown in figure above.

Electrons in Three Dimensions

Energy Levels in 3D

The Schrödinger equation in the three dimensions takes the form

$$H\psi_n(x, y, z) = -\frac{\hbar^2}{2m}\left(\frac{\partial^2}{\partial x^2} + \frac{\partial^2}{\partial y^2} + \frac{\partial^2}{\partial z^2}\right)\psi_n(x, y, z) = E_n\psi_n(x, y, z)$$

If the electrons are confined to a cube of edge L, the solution is the standing wave

$$\psi_n(x, y, z) = A\sin\left(\frac{n\pi}{L}x\right)\sin\left(\frac{n\pi}{L}y\right)\sin\left(\frac{n\pi}{L}z\right)$$

where n_x, n_y, and n_z are positive integers.

In many cases, however, it convenient to introduce periodic boundary conditions, as we did for lattice vibrations. The advantage of this description is that we assume that our crystal is infinite and disregard the influence of the outer boundaries of the crystal on the solution. We require then that our wave function is periodic in x, y, and z directions with period L, so that

$$\psi\ (x+L,\ y,z)\ =\psi\ (x,\ y,z)$$

and similarly for the y and z coordinates. The solution of the Schrödinger equation $H\psi_n(x) = E_n\psi_n(x)$ which satisfies these boundary conditions has the form of the traveling plane wave:

$$\psi_k(\vec{r}) = A\exp\,(i\vec{k}\bullet\vec{r})$$

provided that the component of the wavevector k are determined from

$$k_x = \frac{n\pi_x}{L},\ k_y = \frac{n\pi_y}{L},\ k_z = \frac{n\pi_z}{L}$$

where n_x, n_y, and n_z are positive or negative integers.

Dispersion Relations in 3D

If we now substitute this solution to Eq. $H\psi_n(x) = E_n\psi_n(x)$ we obtain for the energies of the orbital with the wavevector k,

$$E_k = \frac{\hbar^2}{2m}\left(k_x^2 + k_y^2 + k_z^2\right)$$

The dispersion relation is,

$$E_k = \frac{\hbar^2}{2m}k^2$$

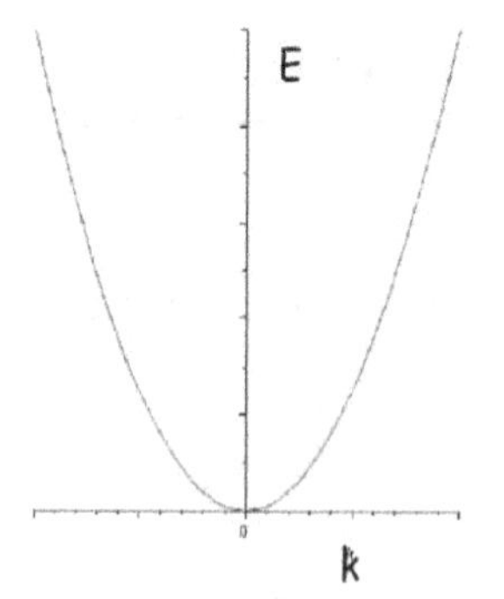

$$E = \frac{\hbar^2 k^2}{2m}$$

Figure: Dispersion relation for 3D.

Density of States in 3D Case

In three dimensional case, wavevector k has discrete values.

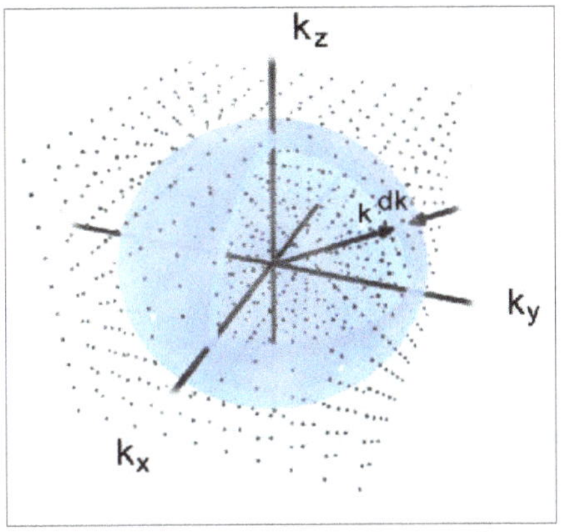

Figure: Visualization of k-space showing values of k as points. The number of allowed states is the number of these points contained in the shell of radius k and thickness dk.

The diagram shows the spherical shells. The problem of finding the number of allowed states amounts to finding the number of these allowed states between spheres of radius k and k +d k . In 3-dimensions, the volume between the two shells is given by:

$$d\vec{k} = dk_x dk_y dk_z = 4\pi k^2\ dk$$

Therefore, the number of states is given simply by dividing this volume by the volume of a single energy state. At this point is convenient to introduce an additional factor of two to account for the intrinsic angular momentum of the electrons or spin states.

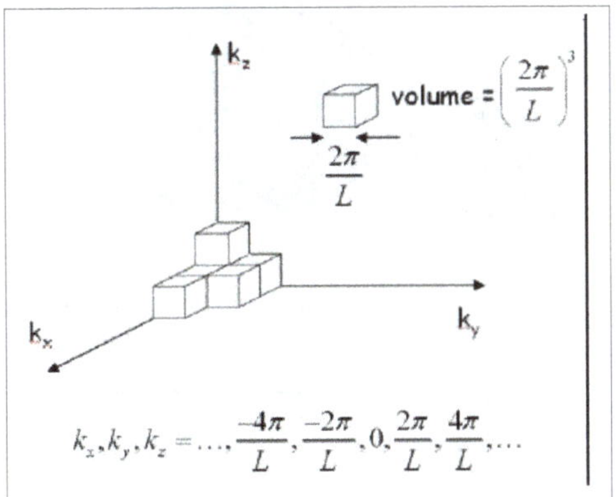

Figure: Discrete k values for 3D

$$k_x = \frac{2\pi n_x}{L},\ k_y = \frac{2\pi n_y}{L},\ k_z = \frac{2\pi n_z}{L}$$

This equation suggests that there is one quantum state within the momentum interval $(2\pi / L)^3$. Consequently the number of quantum states dN in the momentum interval dk is

$$dN = \frac{d\vec{k}}{(2\pi / L)^3} = (\frac{L}{2\pi})^3\ d\vec{k} = (\frac{L}{2\pi})^3\ 4\pi k^2\ dk$$

Multiplying by 2 for spin degeneracy, the number of states within the interval dk,

$$dN = 2\times(\frac{L}{2\pi})^3\ 4\pi k^2\ dk$$

For 3 dimensional quantum case, we have,

$$E = \frac{\hbar^2 k^2}{2m}$$

$$dk = \sqrt{\frac{m}{2\hbar^2 E}}dE$$

Therefore, the density of states per unit length in 1-d and multiplying by 2 for spin degeneracy,

$$dN = 4\pi\times(\frac{L}{2\pi})^3\left(\frac{2m}{\hbar^2}\right)^{3/2} E^{1/2}\ dE$$

For 3 dimension, the density of states per unit volume at energy E is given by

$$DOS = \frac{dN}{dE} = 4\pi\times(\frac{L}{2\pi})^3\left(\frac{2m}{\hbar^2}\right)^{3/2} E^{1/2}$$

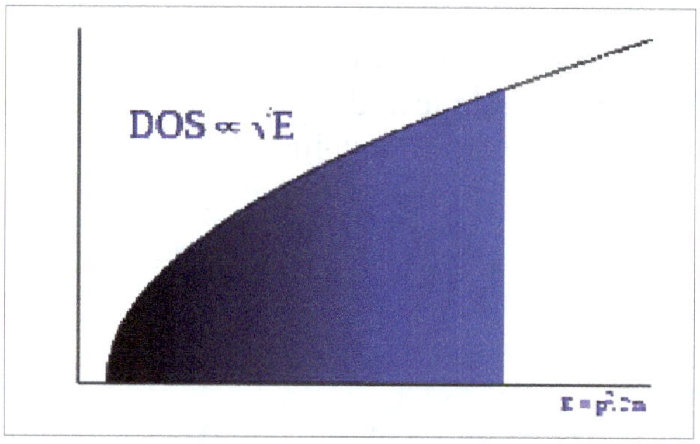

Figure Density of states of three dimensional case.

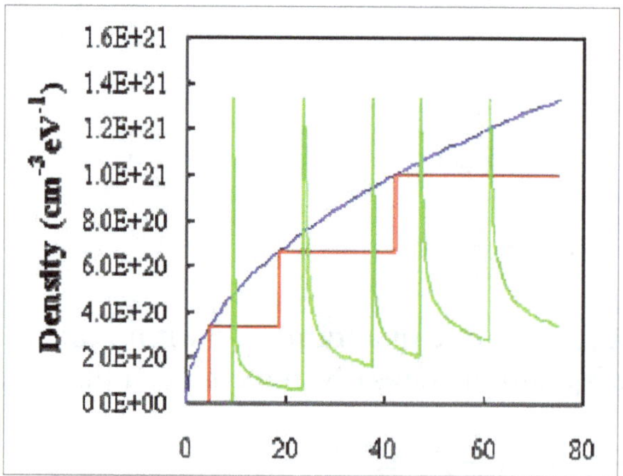

Figure Density of states for bulk (3D blue).

Fermi Energy in 3D case

From the equation of DOS, we have

$$dN = 4\pi \times (\frac{L}{2\pi})^3 \left(\frac{2m}{\hbar^2}\right)^{3/2} E^{1/2} dE$$

$$N = \int dN = \int_0^{\infty} 4\pi \times (\frac{L}{2\pi})^3 \left(\frac{2m}{\hbar^2}\right)^{3/2} E^{1/2} dE = \frac{L^3}{2\pi^2}\left(\frac{2m}{\hbar^2}\right)^{3/2} \int_0^{E_F} E^{1/2} dE = \frac{L^3}{3\pi^2}\left(\frac{2m}{\hbar^2}\right)^{3/2} E_F^{3/2}$$

$$E_F = \frac{\hbar^2}{2m}\left(3\pi^2 \frac{N}{L^3}\right)^{2/3} = \frac{\hbar^2}{2m}\left[\left(3n\pi^2\right)^{1/3}\right]^2,$$

where n is the electron density. Comparison to the equation.

$$E = \frac{\hbar^2}{2m} k^2$$

leads to the Fermi wave vector,

$$k_F = \left(3n\pi^2\right)^{1/3}$$

Because the linear momentum of an electron is,

$$\vec{p} = m\vec{\upsilon} = \hbar\vec{k},$$

We can derive the Fermi velocity for an electron as

$$\upsilon_F = \frac{\hbar k_F}{m} = \frac{\hbar}{m}\left(3n\pi^2\right)^{1/3}$$

An alternative way to understand the Fermi energy is list as below.

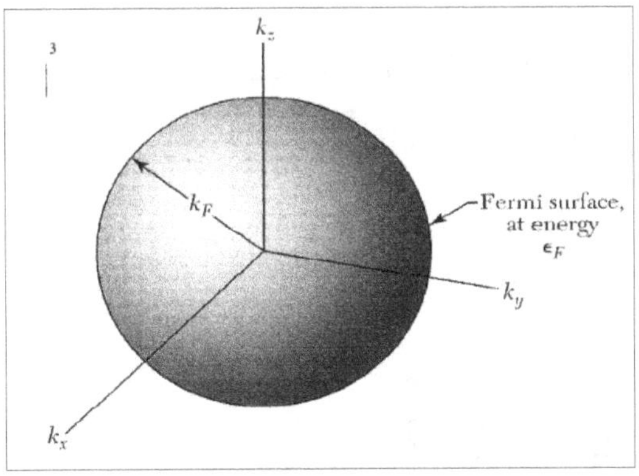

Figure: Fermi surface.

$$N = 2 \times \frac{4\pi k_F^2 / 3}{(2\pi / L)^3}$$

We can get the same expressions of the Fermi wave vector and Fermi energy level.

Example:

Calculate the Fermi energy, Fermi wavevector, Fermi velocity and Fermi temperature for metal Li, whose electron density is $4.70\times10^{22}\ cm^{-3}$.

Solution: $E_F = \frac{\hbar^2}{2m}\left[\left(3n\pi^2\right)^{1/3}\right]^2 = 4.32eV$,

where n is the electron density.

The Fermi wave vector is given by,

$$k_F = \left(3n\pi^2\right)^{1/3} = (3\times4.7\times10^{28}\times3.14^2)^{1/3} = 1.12\times10^{10}\ m^{-1} = 1.12\times10^{8}\ cm^{-1}$$

We can derive the Fermi velocity for an electron as,

$$\upsilon_F = \frac{\hbar k_F}{m} = \frac{1.05\times10^{-34}}{0.91\times10^{-30}}\times1.12\times10^{10} = 1.29\times10^{6}\ m/s$$

The Fermi temperature is given by

$$T_F = E_F / k = 4.32\times1.6\times10^{-19} / 1.38\times10^{-23} = 5.0086\times10^{4}\ K$$

Example: Calculate the number of states per unit energy in a 100 by 100 by 10 nm piece of silicon $\left(m^* = 1.08\ m_0\right)$ 100 meV above the conduction band edge. Write the result in units of eV^{-1}.

Solution: The density of states equals:

$$DOS = \frac{dN}{dE} = (\frac{1}{2\pi^2})\left(\frac{2m}{\hbar^2}\right)^{3/2} E^{1/2}$$

$$DOS = (\frac{8\pi\sqrt{2}}{h^3})\, n*^{3/2}\,(E - E_c)^{1/2} = 1.51\ 10^{56}\ m^{-3}J^{-1}$$

So that the total number of states per unit energy equals:

$$g(E)V = 1.51\times10^{56}\times10^{-22}\ J^{-1} = 2.41\times10^{5}\ eV^{-1}$$

$$DOS = 1.51\times10^{56}\times10^{22}\ J^{-1} = 2.41\times10^{5}\ eV^{-1}$$

Electron Heat Capacity

The question that caused the greatest difficulty in the early development of the electron theory of metals concerns the heat capacity of the conduction electrons. Classical statistical mechanics predicts that a free particle should have a heat capacity of $3/2k_B$, where k_B is the Boltzmann constant. If N atoms each give one valence electron to the electron gas, and the electrons are freely mobile, then the electronic contribution to the heat capacity should be $3/2Nk_B$, just as for the atoms of a monatomic gas. But the observed electronic contribution at room temperature is usually less than 0.01 of this value.

This discrepancy was resolved only upon the discovery of the Pauli exclusion principle and the Fermi distribution function. When we heat the specimen from absolute zero not every electron gains an energy $\sim k_B T$ as expected classically, but only those electrons, which have the energy within an energy range $k_B T$ of the Fermi level, can be excited thermally. These electrons gain an energy, which is itself of the order of $k_B T$. This gives a qualitative solution to the problem of the heat capacity of the conduction electron gas. If N is the total number of electrons, only a fraction of the order of $k_B T$ /EF can be excited thermally at temperature T, because only these lie within an energy range of the order of $k_B T$ of the top of the energy distribution.

Each of these $Nk_B T / E_F$ electrons has a thermal energy of the order of $k_B T$. The total electronic thermal kinetic energy U is of the order of $U \approx (Nk_B T / E_F) k_B T$. The electronic heat capacity is $C_{el} = dU / dT \approx Nk_B (k_B T / E_F)$ and is directly proportional to T, in agreement with the experimental results discussed in the following topic. At room temperature C is smaller than the classical value $\approx Nk_B$ by a factor 0.01 or less.

We now derive a quantitative expression for the electronic heat capacity valid at low temperatures $k_B T << EF$. The total energy of a system of N electrons at temperature T is,

$$U = \int_0^\infty ED(E) f(E,T) dE$$

,

where $f(E,T)$ is the Fermi distribution function and $D(E)$ is the density of states. The heat capacity can be found by differentiating this equation with respect to temperature. Since only the distribution function depends on temperature we obtain:

$$C_{el} = \frac{dU}{dT} = \int_0^\infty ED(E) \frac{df(E,T)}{dT} dE.$$

It is more convenient to represent this result in a different form:

$$C_{el} = \int_0^\infty (E - E_F) D(E) \frac{df(E,T)}{dT} dE.$$

Equation $0 = E_F \frac{dN}{dT} = E_F \int_0^\infty D(E) \frac{dF(E,T)}{dT} dE$ .is equivalent to Equation $C_{el} = \frac{dU}{dT} = \int_0^\infty ED(E) \frac{df(E,T)}{dT} dE$. due to the fact which follows from equation $N = \int_0^{E_F} D(E) dE$,:

$$0 = E_F \frac{dN}{dT} = E_F \int_0^\infty D(E) \frac{dF(E,T)}{dT} dE$$

Since we are interested only temperatures for which $k_B T << E_F$ the derivative df / dT is large only at the energies which lie very close to the Fermi energy. Therefore, we can ignore the variation of D(E) under the integral and take it outside the integrand at the Fermi energy, so that

$$C_{el} = D(E_F)\int_0^{\infty}(E-E_F)\frac{df(E,T)}{dT}\,dE\,.$$

We also ignore the variation of the chemical potential with temperature and assume that $\mu = E_F$, which is good approximation at room temperature and below. Then

$$\frac{df(E,T)}{dT} = \frac{\mathrm{E}-E_F}{k_B T^2}\frac{e^{(E-E_F)k_BT}}{\left[e^{(E-E_F)k_BT}+1\right]^2}\,.$$

Equation $\frac{df(E,T)}{dT} = \frac{\mathrm{E}-E_F}{k_B T^2}\frac{e^{(E-E_F)k_BT}}{\left[e^{(E-E_F)k_BT}+1\right]^2}$ .can, then, be rewritten as,

$$C_{el} = D(E_F)\int_0^{\infty}\frac{(E-E_F)^2}{k_B T^2}\frac{e^{(E-E_F)k_BT}}{\left[e^{(E-E_F)k_BT}+1\right]^2}dE = D(E_F)\int_{-E_F/k_BT}^{\infty}\frac{x^2(k_BT)^3}{k_BT^2}\frac{e^x}{\left(e^x+1\right)^2}dx\,.$$

Taking into account that $E_F >> k_B T$, we can put the low integration limit to minus infinity and obtain,

$$C_{el} = D(E_F)k_B^2 T\int_{-\infty}^{\infty}\frac{x^2 e^x}{\left(e^x+1\right)^2}dx = \frac{\pi^3}{3}D(E_F)k_B^2 T$$

For a free electron gas we should use equation $D(E)=\frac{3N}{2E}$ for the density of states to finally obtain,

$$C_{el} = \frac{\pi^2}{2}Nk_B\frac{T}{T_F},$$

where we defined the Fermi temperature $T_F = E_F/k_B$. This is similar to what we expected to obtain according to the qualitative arguments given in the beginning of this topic.

Experimentally the heat capacity at temperatures much below both the Debye temperature and the Fermi temperature can be represented in the form:

$$C = C_{el} + C_{ph} = \alpha T + \beta T^3$$

The electronic term is dominant at sufficiently low temperatures. The constants α and β can be obtained by fitting the experimental data.

Fermi Energy

Fermi energy is often defined as the highest occupied energy level of a material at absolute zero temperature. In other words, all electrons in a body occupy energy states at or below that body's Fermi energy at oK.

The fermi energy is the difference in energy, mostly kinetic. In metals this means that it gives us the velecity of the electrons during conduction. So during the conduction process, only electrons that have an energy that is close to that of the fermi energy can be involved in the process.

This concept of Fermi energy is useful for describing and comparing the behaviour of different semiconductors. For example: an n-type semiconductor will have a Fermi energy close to the conduction band, whereas a p-type semiconductor will have a Fermi energy close to the valence band.

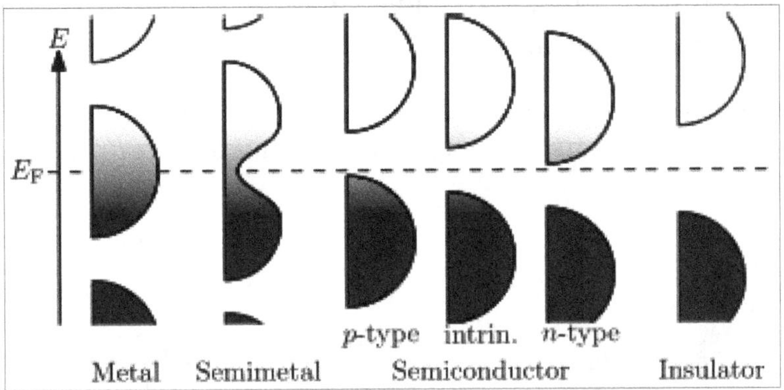

Fermi energies of different material types.

As a material's temperature rises above absolute zero, the probability of electrons existing in an energy state greater than the Fermi energy increases, and there is no longer any constant highest occupied level, so while the material's Fermi energy may be useful as a reference, it is not very useful at real temperatures.

Instead, we can approximate the *average* energy level at which an electron is present is with the Fermi-Dirac distribution:

$$f(E) = \frac{1}{1 + e^{\frac{E - E_F}{kT}}}$$

where E is the energy level, k is the Boltzmann constant, T is the (absolute) temperature, and E_F is the Fermi level. The Fermi level is defined as the chemical potential of electrons, as well as the (hypothetical) energy level where the probability of an electron being present is 50%.

You can calculate the fermi energy state using:

$$E_f = \left(\frac{h^2}{8m}\right)\left(\frac{3N}{\pi V}\right)^{\frac{2}{3}}$$

N - number of possible quantum states

V - volume

m - mass of electron

h - planc's constant

Fermi Surface

Fermi surface is the surface in k space that corresponds to the Fermi energy. In quantum mechanics we accept the wave particle duality and associate particles with a waves consistent with their momentum. In k space, we are plotting wave vectors $k = \frac{2\pi}{\lambda}$, where λ is the wavelength corresponding to electrons with specific energy. Therefore, the Fermi surface represents wave vectors of electrons having the Fermi energy E_f. To understand the significance of this, let us briefly look at the interaction of radiation with matter.

When radiation in the form of X-rays or an electron beam, interacts with matter, we see diffraction effects. Figure below shows a schematic of a typical experiment to record diffraction patterns.

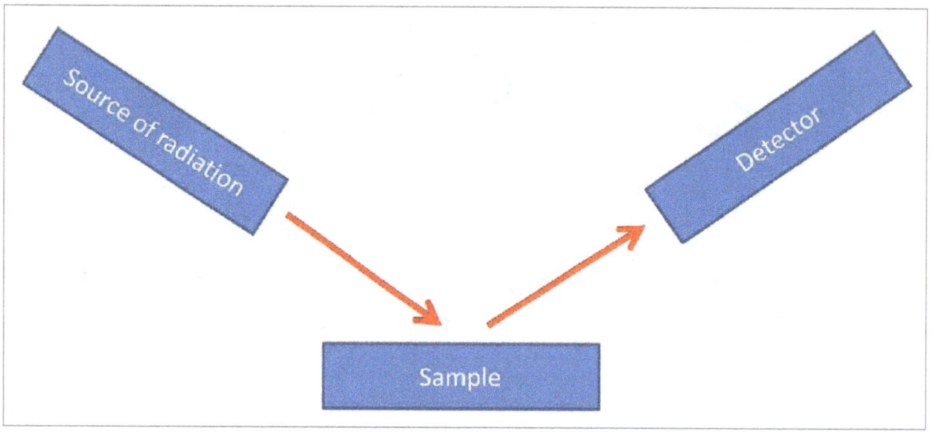

Figure: A schematic showing how radiation can be made to interact with matter to produce and record diffraction effects in a laboratory setting.

To enable recording diffraction effects, in general we need:

1) A source of radiation
2) Sample with a periodic structure
3) Wavelength of the radiation to be in a range that enables diffraction effects to be recorded (which is of the order of the spacing of the periodic structure, or less)

However, the important aspect to note is that even though the schematic shows the source of the radiation to be independent of the sample, this is only an experimental convenience. Waves of electrons already within the sample can interact with the periodic structure of the sample and display diffraction effects. Unlike a separate electron beam source in an electron microscope, this is the equivalent of an electron beam within the sample.

The Fermi surface represents the wave vectors of the electrons corresponding to the Fermi energy. k space plots this information in inverse length dimensions. We can plot the periodic structure of the material also using inverse length notation, on the same plot. This combined plot can help us identify diffraction conditions for the nearly free electrons within the material due to interaction with the periodic structure of the same material. The plot of the periodic structure of the material in inverse length dimensions is referred to as a plot in 'Reciprocal space' - which will be examined in greater detail in a subsequent class, and is a very important tool in understanding the behavior of electrons in solids. For the moment we will simply state in summary that the diffraction of nearly free electrons results in the formation of permitted energy levels and forbidden energy levels, or bands for the free electrons.

Fermi Temperature

The Fermi temperature T_f, is defined as the temperature corresponding to the Fermi energy such that:

$$E_f = k_b T_f$$

With the Fermi energy of the order of a few eV, the Fermi temperature works out to be of the order of several thousand Kelvin (approximately 10,000 K). How do we reconcile the fact that the material is at room temperature or lower, or even at 0 Kelvin, while the Fermi temperature is 10,000 K? One way to look at the situation is as follows: When we measure the temperature of a material, we do not typically measure the temperature of a single atom or electron. What we measure is the average temperature of the material. There is invariably going to be a distribution of energy within the material. In this distribution, an extremely small thermal mass, consisting of a very small fraction of the nearly free electrons (which is itself a very small fraction of the total electrons in the system), is at the Fermi energy, and the temperature corresponding to that energy is the relatively high Fermi temperature. Therefore the 'high' Fermi temperature is not inconsistent with the 'low' temperature or the solid as a whole.

5

Classical System of Interacting Particles

The classical system of interacting particles can be explained by cluster expansion and virial coefficients. Cluster expansion gives a method to understand the logarithm of partition function. In order to completely understand the classical system of interacting particles, it is necessary to understand the processes related to it. The following chapter elucidates the varied processes and mechanisms associated with this area of study.

In an ideal gas system the density of the system is very low that particles are so far apart that interaction among the particles is negligible. We wish to look at the real systems or the so called non-ideal gases restricting to classical systems only. As the density increases the particles start coming closer to each other and interactions no longer remain negligible and ideal gas equation of state no longer remains valid. Therefore, our main focus will be to see how ideal gas equation of state gets modified as a result of such an interaction. In fact number of empirical and semi-empirical equations of states have been proposed for non-ideal gases. The most famous being Van der Waals equation of state, after the name of Dutch physicist Van der Waal, which met with a reasonable success to explain qualitatively the behavior of the real gas system near the gas liquid critical point.

Later Theisen proposed and Kamerlingh Onnes carried the study forward providing a very strong theoretical background and proposed the virial equation of state as an infinite power series in terms of number density $\frac{n}{v}$,

$$\frac{P}{k_B T}=\frac{n}{V}+B_2(T)\left(\frac{n}{V}\right)^2+B_3(T)\left(\frac{n}{v}\right)^3+\ldots$$

The quantities $B_2(T), B_3(T)$.... are called the second, third,....virial coefficients respectively which depend only on temperature and are independent of pressure and density. The jth virial coefficient in fact corresponds to inclusion of j particle interactions in a given volume V.

To prepare for this derivation, we first have a look at some possible interaction energy forms or so called model potentials. This will be followed by detailed study of a hard sphere gas to calculate the partition function in terms of configurational integral solved for 1- dimensional case and further generalized to higher dimensions by method of induction to obtain Free energy. This will be supplemented further by introducing weak attractive interaction to calculate, by perturbation method, free energy. The two results together shall be used to derive well known Van der Waals equation of state for a real gas whose isotherms shall be plotted to identify its limitations and remedy these by introducing the well-known Maxwell's construction and understanding of gas liquid transition.

Model Potentials

Here number of model potentials which were proposed from time to time which were supposed to take care of two broad features expected in the real behavior of atoms in gases. At very short distances, potentials should be such that atoms at such distances should not sit on top of each other demanding a repulsive part and at moderate distances should allow for keeping the atoms together, ultimately going to zero at large distances. In the following model potentials we will see this reality being introduced. Out of these model potentials because of some very empirical reasons some potentials emerged to be favourites for making theoretical studies of interacting classical systems. Let us have a look at few of these.

Hypothetical Model Potential

A typical hypothetical inter particle potential V(r) is given in the plot below:

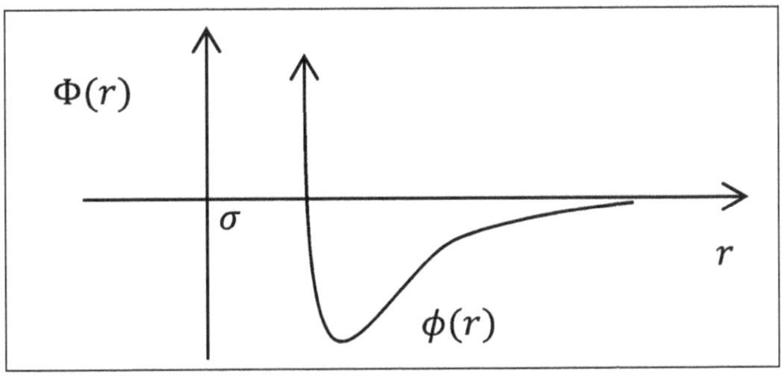

It is showing a strongly repulsive part with a weak attractive well, its shape depends upon the structure of the molecules, r is the intermolecular distance.

Hard Sphere Potential

The hard sphere potential is represented mathematically as:

$$\phi(r)=\begin{cases}\infty \text{ for } r<\sigma \\ 0 \text{ for } r>\sigma\end{cases}$$

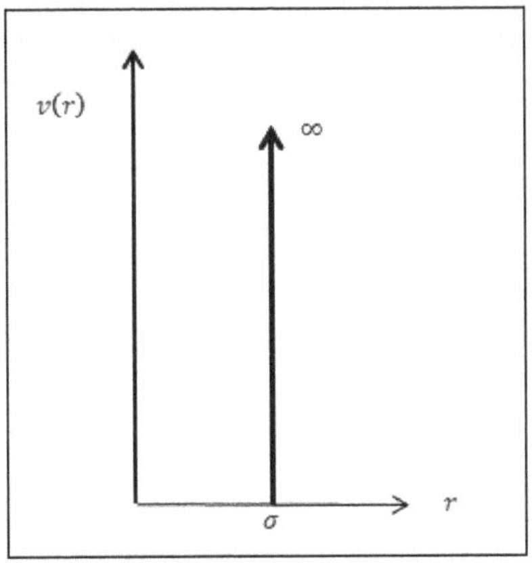

Graphically, it is shown in figure. This potential has repulsive part only and no attractive part, yet very closely represents the steep repulsive parts of realistic potentials. It is the simplest potential and there for analytically tractable to calculate virial coefficients upto 7th order. A gas modeled on this potential is called hard sphere gas or fluid.

Square Well Potential

Square well potential is an improvement over hard sphere potential with the inclusion of an attractive part keeping it analytically tractable. It has the form

$$\phi(r)=\begin{cases}\infty \text{ for } r<\sigma \\ -\varepsilon\sigma<r<\lambda\sigma \\ 0 \text{ for } r>\lambda\sigma\end{cases}$$

Where λ is the range of attractive potential whose depth is $-\varepsilon$. Graphically it can be visualised as given in figure below:

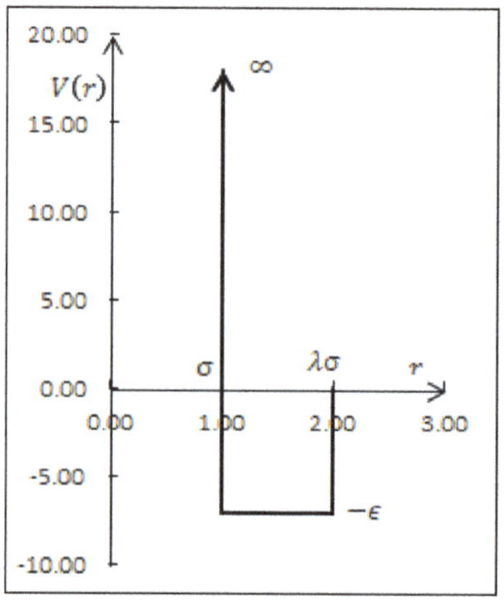

Square well potential with range σ and depth ϵ.

Lenard Jones Potential

The hard sphere and square well potentials are nowhere closer to the hypothetical potential and are very far from actual inter particle potential. Lenard Jones potential is the closest to the hypothetical potential which qualitatively represents the desirable behavior. It has an attractive part with an appropriately incorporated short distance repulsive part. It is also called 6-12 potential and algebraically has the following form:

$$\phi(r)=4\varepsilon\left(\left(\frac{\sigma}{r}\right)^{6}-\left(\frac{\sigma}{r}\right)^{12}\right)$$

It has a minimum at r = σ. In fact it has been chosen out of a generic potential called n-m potential represented by the form.

$$\phi(r)=4\varepsilon\left(\left(\frac{\sigma}{r}\right)^{n}-\left(\frac{\sigma}{r}\right)^{m}\right)$$

Where m < n, after it was found that n=6 and m=12 represents the best approximation for most of the cases. Also it must be emphasized that it is a two body potential and neglects three body and higher number of bodies interactions. The attractive part is a result of each particle (atom or molecule) treated as a dipole resulting in an attractive potential with power 6. This potential indeed represents the Van der Waal's force between the two particles. r^{12} repulsive term is a result of Pauli's repulsion at short ranges due to overlapping atomic orbitals. 6-12 potential is plotted in figure below:

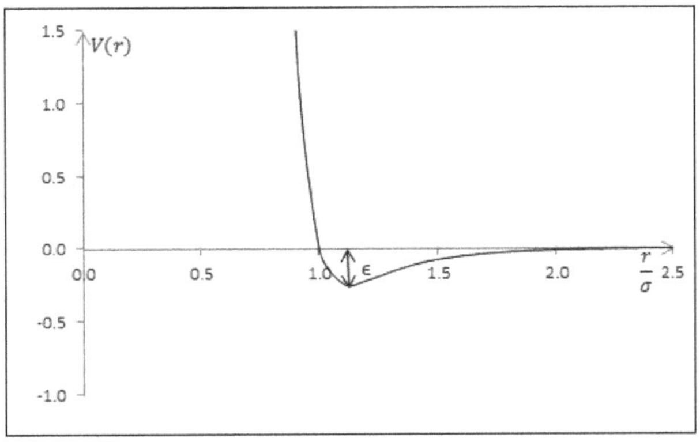

Plot of Lennard Jones potential, at r = σV(r) = 0 and minima of the potential occurs at $r = 2^{\frac{1}{6}}\sigma$, depth of the minima is ϵ.

Lennard Jones potential has been used extensively to study many properties of solids, liquids and gases successfully.

Hard Sphere Gas

As stated earlier, a hard sphere gas is a system in which a gas is modelled on hard sphere potential. It implies that gas is made up of particles (atoms or molecules), such that their radius is unaffected by any external influence such as change in temperature, pressure or collision among each other. The closest distance, σ, the two particles can come is, when the distance between the centres of two hard spheres is equal to the diameter of the sphere i.e. 2 times the radius of the sphere (r_0), as shown in figure below:

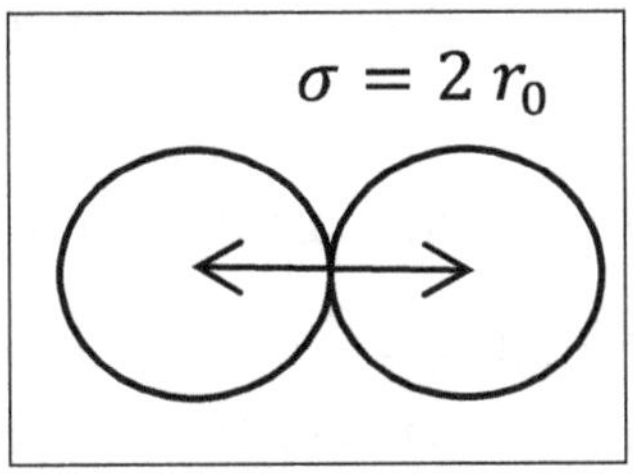

Closest distance between two hard spheres.

Also from the plot of the hard sphere potential it is clear that for distance greater than σ there is no interaction between the particles.

The Hamiltonian and Configurational Integral

Now denoting the position of the centres of spheres constituting the system by r_i, where i goes from 1 to N, the Hamiltonian of the system can be written as,

$$H = H_{ideal} + \phi(r_1, .., r_2)$$

Where $H_{ideal} = \Sigma_i \frac{p_i^2}{2m}$ is the Hamiltonian corresponding to the ideal gas and $V(r_1,..,r_2)$ is the hard sphere potential energy given by,

$$\phi(r_1,...,r_2) = \begin{cases} \infty \text{ for any } |r_i - r_j| < \sigma \\ 0 \text{ for } |r_i - r_j| > \sigma \end{cases}$$

Since we are assuming only two particle interaction only, above equation can also be written as,

$$\phi(r_1,...,r_2) \sum_i^{i>j} \sum_j \phi_2(|r_i - r_j|)$$

where,

$$\phi_2(|r_i - r_j|) = \begin{cases} \infty \text{ for } |r_i - r_j| < \sigma \\ 0 \text{ for } |r_i - r_j| > \sigma \end{cases}$$

With the Hamiltonian given by $(H = H_{ideal} + \phi(r_1,..,r_2))$, the partition function of system can be written as,

$$Q_N(V,T) = \frac{1}{N!h^{DN}} \int e^{\left\{-\beta \Sigma_i \frac{p_i^2}{2m} - \beta \Sigma_i^{i>j} \Sigma_j \phi_2(|r_i - r_j|)\right\}} d^{DN} p \, d^{DN} r$$

Here N is the number of particles and D is the dimension of the system. Factor N! comes because of in distinguishability of the particles in the system. Obviously now above equation can be written as,

$$Q_N(V,T) = \frac{1}{N!h^{DN}} \int e^{\left\{-\beta \Sigma_i \frac{p_i^2}{2m}\right\}} d^{DN} p \int e^{\left\{-\beta \Sigma_i^{i>j} \Sigma_j \phi_2(|r_i - r_j|)\right\}} d^{DN} r$$

Decoupling the momentum part and position part and hence momentum integral can be solved easily given by $(2\pi mkT)^{\frac{DN}{2}}$. Recalling that $\frac{h}{(2\pi mkT)^{\frac{1}{2}}} = \lambda$, mean thermal de Broglie wave length,

$\left(Q_N(V,T) = \frac{1}{N!h^{DN}} \int e^{\left\{-\beta \Sigma_i \frac{p_i^2}{2m}\right\}} d^{DN} p \int e^{\left\{-\beta \Sigma_i^{i>j} \Sigma_j \phi_2(|r_i - r_j|)\right\}} d^{DN} r\right)$ can be written as,

$$Q_N(V,T) = \frac{1}{N!\lambda^{DN}} \int e^{\left\{-\beta \Sigma_i^{i>j} \Sigma_j \phi_2(|r_i - r_j|)\right\}} d^{DN} r.$$

Here the remaining integral is over the space co-ordinates of the N particles only, and is configurational integral, represented by Z_N (V,T) such that,

$$Z_n(V,T) = \int e^{\left\{-\beta \Sigma_i^{i>j} \Sigma_j \phi_2(|r_i - r_j|)\right\}} d^{DN} r$$

For the ideal gas, $\phi_2\left(\left|r_i - r_j\right|\right) = 0$, so that $Z_N^{ideal}(V,T) = V^N$ and $Q_N^{ideal}(V,T) = \frac{V^N}{N!\lambda^{DN}}$ as is the case. In the case of interacting (non-ideal) systems, main task is to calculate the configurational integral.

One Dimensional Hard Sphere Gas Exact Solution

Formulation of a one dimensional hard sphere gas, containing N particles confined linearly over a length L, can be visualized as placing spherical beads of radius r_0 on a wire of length L.

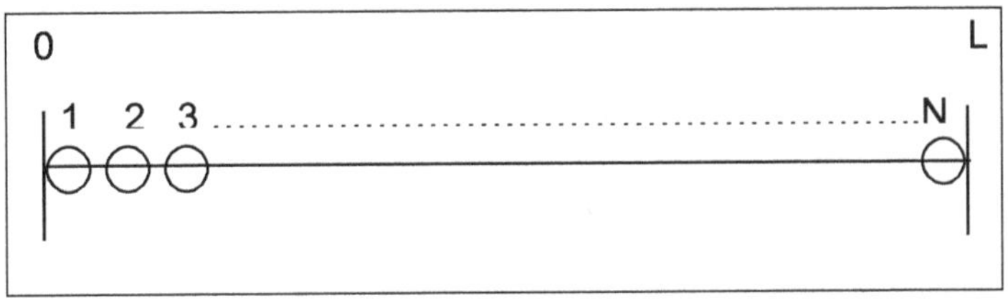

A hard sphere gas in one dimension, each bead is of radius r_0 distributed over a wire of length L.

Note that from the either edge of the wire a bead cannot be positioned with its centre at a distance less than the radius r_0 of the bead. Secondly, the length L of the wire must be greater than $N(2r_0)$. So the problem of evaluating configurational integral can built up inductively as follows:

For a Hard sphere gas containing only one sphere confined over a length L, configurational integral can be written as,

$$Z_1(L,T) = \int_{r_0}^{L-r_0} dx_1 = x_1\Big|_{r_0}^{L-r_0} = (L - 2r_0)$$

We can write this alternately, suppose the wire's left end is positioned at a and right end at b, then length of the wire L = (b – a) and, therefore above equation can be written as,

$$Z_1(L,T) = Z_1(a,b,T) = (b - a - 2r_0)$$

To proceed further, we state an ansatz, that for distributing N particles over the wire of length L, the configurational integral Z_N (L, T) is given by,

$$Z_N(a,b,T) = (b - a - 2r_0N)^N$$

and to prove the above assumption, we apply method of induction, according to which if we take N = 1, it is obviously true, and if it can be proved true for N = N + 1, the stated ansatz can be established as true.

Proving this amounts to adjusting one more particle on the wire, let us assume that it is positioned at (k + 1) th position denoted by x, such that there are k particles towards left and N – k particles towards right, where k can take any integer value between 0 and N. The net configurational integral for adjusting (k + 1)th particle can be written as a sum over all the configurational integrals with each term weighted by the factor counting the possible ways of selecting k particles towards left of (k + 1) particle out of a total of N particles, given by (NK). That is you are putting k particles over a length bounded by (0, x – r_0) and N – K particles over a length bounded by (x + r_0, L).

Thus we can write,

$$Z_{N+1}=\sum_{k=o}^{N}\binom{N}{k}Z_k\left(o,x-r_o\right)\int_{x_{max}}^{x_{max}}Z_{n-k}\left(x+r_{o,}L\right)dx.$$

For the sake of simplicity we have omitted T. Using the ansatz ($Z_N\left(a,b,T\right)=\left(b-a-2r_oN\right)^N$), we can rewrite ($Z_{N+1}=\sum_{k=o}^{N}\binom{N}{k}Z_k\left(o,x-r_o\right)\int_{x_{max}}^{x_{max}}Z_{n-k}\left(x+r_{o,}L\right)dx.$) as,

$$Z_{N+1}=\int_{x_{mini}}^{x_{max}}\left[\sum_{K=o}^{N}\binom{N}{K}\left(x-\left(2k+1\right)r_o\right)^k\left(L-x-\left(2\left(N-k\right)-1\right)\right)^{N-k}\right]dx$$

Recalling, binomial theorem for a positive integer, $\left(x+a\right)^N=\sum_{k=o}^{N}\binom{N}{k}x^ka^{N-k}$, above equation can be written as,

$$Z_{N+1}=\int_{x_{min}}^{x_{max}}\left(L-2\left(N+1\right)r_o\right)^N dx=\left(L-2\left(N+1\right)r_o\right)^N\int_{x_{min}}^{x_{max}}dx$$

Now let us look at the limits of x, k particles are distributed over a distance k ($2r_o$) + r_o so that x_{min} = (2k + 1) r_o and N - k particles must be distributed over a distance to accommodate N – k particles (N - k) × 2 r_o + r_o = (2(N - k) + 1) r_o, so that x_{max} = L – (2(N – k) + 1) r_o. Using these limits the integral yields x_{max} - x_{min} = L – 2(N + 1) r_o we find that,

$$Z_{N+1}=\left(L-2\left(N+1\right)r_o\right)^{N+1}$$

Proving the ansatz.

Thus configurational integral for distributing N particles each of radius r_o is,

$$Z_N=\left(L-2\left(N\right)r_o\right)^N$$

Provided N is an integer $<\frac{L}{2r_o}$.

And the partition function is given by,

$$Q_N\left(L,T\right)=\frac{\left(L-2\left(N\right)r_o\right)^N}{N!\lambda^N}$$

The free energy of the one dimensional hard sphere can be obtained as,

$$A=-k_BT\,In\,Q_N\left(L,T\right)=-Nk_BT\left[In\frac{\left(L-2Nr_o\right)}{N\lambda}+1\right]$$

Generalisation to Higher Dimensions and Thermodynamic Properties

Let us look at some interesting features of the configurational integral, (21), derived for 1-dimensional hard sphere gas:

- For r_o = 0, configurational integral reduces to LN.

- Effective length available is L-2(N) $r_{o.}$

This suggests that for a higher dimensional case, DN=3, say L should get replaced by V^N for r_o = o and for a hard sphere gas should lead to an effective volume V, should get replaced by V – bN, where b has the dimensions of volume and is a function of the radius r_o of the sphere.

Thus for a system in higher dimensions, the configurational integral can be written as,

$$Z_N = (V - Nb)^N$$

For r_o = $_o$, b=o and, therefore, configurational integral reduces to V^N.

Hence partition function for the hard sphere comes out to be

$$Q_N(V,T) = \frac{(V-Nb)^N}{N!\lambda^N}$$

now we can obtain free energy,

$$A = -k_B T \operatorname{In} Q_N(V,T) = -NK_B T\left[\operatorname{In}\frac{(V-Nb)}{N\lambda} + 1\right]$$

putting V=o, yields the same result which we obtain for an ideal gas. Nb is the consequence of hard sphere potential, since because of finite size of the spheres, all the volume is not accessible to the hard spheres. The closest distance to which the centers of two spheres can come, cannot be less than $2r_o$.

A Gas with Attractive Interaction of Finite Range

Let us move from a hard sphere gas to a gas among whose particles there is a finite range weak two body interaction potential:

$$\phi(r_1,\ldots,r_2) = \begin{cases} (-\varepsilon, o) \text{ for } |r_i - r_j| < \sigma \\ o \text{ for } |r_i - r_j| > \sigma \end{cases}$$

where σ is the range of interaction. The Hamiltonian of this problem can be written as,

$$H = H^{id} + H_1$$

where $H_1 = \sum_i^{i>j}\sum_j \phi_2(|r_i - r_j|)$ when $\phi_2(|r_i - r_j|) = o$, the system reduces to an ideal gas.

We shall solve this problem using classical perturbation theory. First we briefly review the perturbation procedure and then apply it to the problem in hand.

Classical Perturbation Theory: Calculation of Partition Function

It is a well known theory in classical/quantum mechanics. This can be applied if Hamiltonian of a system can be written as sum of two terms, with first term usually denoted by H_o and the second term H_1 a very small term compared to H_o called perturbation i.e.

$$H = H_o + H_1, H_o \gg H_1$$

Then if the partition function can be evaluated exactly for H_o, further correction to the partition function because of perturbation H_1 can be evaluated systematically by what is called perturbation expansion procedure.

We rewrite above equation by making smallness of H_1 explicit by introducing a small parameter λ such that,

$$H = H_o + \lambda H_1'.$$

Perturbation expansion is indeed a power series in terms of this parameter *λ* which gives exactly soluble part when put equal to zero.

Now for the partition function we can write, In Z as a power series,

$$\text{In} Z = \text{In} Z_o + \lambda\left(\frac{d\,\text{In} Z}{d\lambda}\right)_{\lambda=o} + \frac{\lambda^2}{Z}\left(\frac{d^2\,\text{In} Z}{d\lambda^2}\right)_{\lambda=o} + \ldots$$

once again it is empahsied, when λ= o, we get partition function for the unperturbed part of the Hamiltonian, H_o this is the part which can be evaluated exactly and is a requisite for the perturbation expansion to be applicable.

Using the notation of quantum mechanics, to apply ($H = H_o + H_1, H_o \gg H_1$), let us look at the partition function,

$$Z = \text{Tr}\left(e^{-\beta H}\right) = \text{Tr}\left(e^{-\beta\left(H_o + \lambda H_1'\right)}\right)$$

Classically trace means integration over whole of phase space divided by suitable power of $2\pi h$.

Using,

$$\frac{d\,\text{In} Z}{d\lambda} = \frac{1}{Z}\frac{dZ}{d\lambda} = \frac{1}{Z}\left(\frac{d\,\text{Tr}\left(e^{-\beta\left(H_o + \lambda H_1'\right)}\right)}{d\lambda}\right) = \frac{1}{Z}\left(\text{Tr}\left(-\beta H_1' e^{-\beta\left(H_o + \lambda H_1'\right)}\right)\right)$$

Recalling, the definition of expectation value,

$$\langle o \rangle = \frac{\text{Tr}\left(Oe^{-\beta H}\right)}{\text{Tr}\left(e^{-\beta H}\right)} = \frac{\text{Tr}\left(Oe^{-\beta H}\right)}{Z}$$

($\text{In} Z = \text{In} Z_o + \lambda\left(\frac{d\,\text{In} Z}{d\lambda}\right)_{\lambda=o} + \frac{\lambda^2}{Z}\left(\frac{d^2\,\text{In} Z}{d\lambda^2}\right)_{\lambda=o} + \ldots$), gives,

$$\frac{d\,\mathrm{In}\,Z}{d\lambda}=\langle -\beta H_1' \rangle = -\beta\langle H_1' \rangle$$

Since we require $\left(\frac{d\,\mathrm{In}\,Z}{d\lambda}\right)_{\lambda=0}$ i.e. derivative evaluated at $\lambda = 0$, the expectation value has to be evaluated with $\lambda = 0$, i.e. with the Hamiltonian H_0 such that,

$$\left(\frac{d\,\mathrm{In}\,Z}{d\lambda}\right)_{\lambda=0} = -\beta\langle H_1' \rangle_0 .$$

Similarly, we can evaluate the second order derivative:

$$\frac{d^2\,\mathrm{In}\,Z}{d\lambda^2}=\frac{d}{dZ}\left(\frac{1}{Z}\left(\frac{dZ}{d\lambda}\right)\right)=\frac{1}{z}\left(\frac{d^2Z}{d\lambda^2}\right)-\left(\frac{1}{Z}\left(\frac{dZ}{d\lambda}\right)\right)^2$$

$$=\frac{1}{Z}\left(\mathrm{Tr}\left(\beta^2 H_1'^2 e^{-\beta(H_0+\lambda H'_1)}\right)\right)-\left(\frac{1}{Z}\mathrm{Tr}\left(-\beta H_1' e^{-\beta(H_0+\lambda H'_1)}\right)\right)^2$$

$$=\beta^2\langle H_1'^2\rangle-\beta^2\langle H_1'\rangle^2$$

Once again, since we require $\left(\frac{d^2\,\mathrm{In}\,Z}{d\lambda^2}\right)_{\lambda=0}$ i.e. derivative evaluated at $\lambda = 0$, the expectation value has to be evaluated with $\lambda = 0$, i.e. with the Hamiltonian H_0 such that,

$$\left(\frac{d^2\,\mathrm{In}\,Z}{d\lambda^2}\right)_{\lambda=0}=\beta^2\left(\langle H_1'^2\rangle_0-\langle H_1'\rangle_0^{\,2}\right)$$

Thus up to second order correction, we have

$$\mathrm{In}\,Z=\mathrm{In}Z_0-\lambda\beta\langle H_1'\rangle_0+\frac{\lambda^2}{2}\beta^2\left(\langle H_1'^2\rangle_0-\langle H_1'\rangle_0^2\right)+\ldots$$

Recalling $H_1=\lambda H_1'$,

Or

$$\mathrm{In}\,Z=\mathrm{In}\,Z_0-\beta\langle H_1\rangle_0+\frac{\beta^2}{2}\left(\langle H_1^2\rangle_0-\langle H_1\rangle_0^2\right)+\ldots$$

Similarly we can evaluate the higher order terms.

First Order Correction to Free Energy

Now we are ready to evaluate first order correction to partition function of an ideal gas, perturbed by a very weak finite attractive interaction defined by equation $H=H^{id}+H_1$.

Thus,

$$\langle H_1\rangle_0=\frac{1}{V^N\int e^{-BH_0}d^{3N}p}\sum_i^{t>j}\sum_j\int d^{3N}p\int d^{3N}r\phi_2\left(|r_i-r_j|\right)e^{-\beta H_0}$$

It is noted that momentum integration in the summation can be taken out which cancels with the momentum integration in the denominator giving,

$$\langle H_1 \rangle_o = \frac{1}{V^N} \sum_i^{t>j} \sum_j \int d^{3N} r \phi_2 \left(|r_i - r_j| \right)$$

Summing over i and j gives terms which are $\frac{N(N-1)}{2}$ in number and give the same contribution, therefore, above equation can be written as,

$$\langle H_1 \rangle_o = \frac{N(N-1)}{2V^N} \int d^{3N} r \phi_2 \left(|r_i - r_j| \right)$$

$$= \frac{N(N-1)}{2V^N} V^{N-2} \int d^3 r_1 \int d^3 r_2 \phi_2 \left(|r_1 - r_2| \right)$$

Where V^{N-2}factor comes from all other co-ordinates except r_1 and r_2. Put $r = r_1 - r_2$, for fixed r_1, $d_r = dr_2$

Hence,

$$\langle H_1 \rangle_o = \frac{N(N-1)}{2V^2} \int 4\pi r_1^2 dr_1 \int 4\pi r_2^2 dr_2 \phi_2 \left(|r_1 - r_2| \right)$$

$$= \frac{4\pi N(N-1)}{2V^2} V \int dr_2 \phi_2 (r)$$

For N large, this can be written as,

$$\langle H_1 \rangle_o = \frac{N^2}{V} 2\pi \int_0^\infty r_2^2 dr_2 \phi_2 (r)$$

Let us look at the integral in equation below since $\phi(r_2)$ as defined in $\phi(r_1,...,r_2) = \begin{cases} (-\varepsilon, 0) \text{ for } |r_i - r_j| < \sigma \\ 0 \text{ for } |r_i - r_j| > \sigma \end{cases}$

is negative and finite, integral must be negative say $-\frac{a}{2\pi}$ i.e.

$$\langle H_1 \rangle_o = -\frac{aN^2}{V}.$$

So the Free energy for the perturbed ideal gas by an attractive potential can be written as,

$$A = -K_B T \ln Q_N (V,T) = -Nk_B T \left[\ln \left(\frac{V}{N\lambda} \right) + 1 \right] - \frac{aN^2}{V}$$

Vander Waal's Equation of State

Now with the treatment of the ideal gas in the presence of repulsive 3-dimensional hard sphere potential and a weak attractive potential, we can go ahead with the discussion of bringing these two together which is the case in real systems by combining together the attractive and repulsive part contributions to the free energy as given in $A=-k_B T \ln Q_N(V,T)=-NK_B T\left[\ln\frac{(V-Nb)}{N\lambda}+1\right]$ and ($A=-K_B T \ln Q_N(V,T)=-Nk_B T\left[\ln\left(\frac{V}{N\lambda}\right)+1\right]-\frac{aN^2}{V}$) the effective potential shown in figure below:

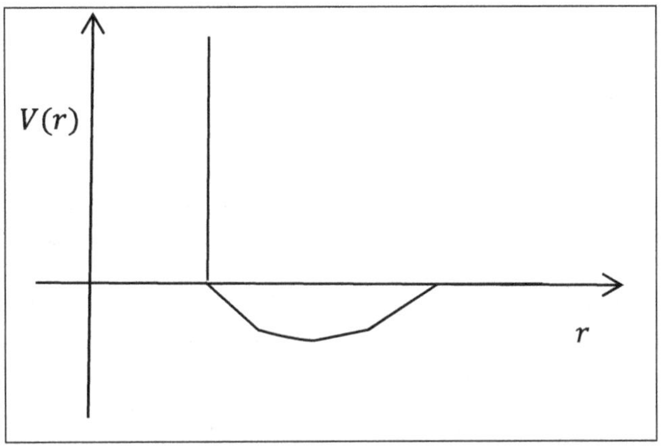

Schematic representation of hard sphere potential with weak attractive part.

$$A=-K_B T \ln Q_N(V,T)=-Nk_B T\left[\ln\left(\frac{V-bn}{N\lambda}\right)+1\right]-\frac{aN^2}{V}$$

We can derive expression for the pressure P from above equation

$$P=-\left(\frac{\partial A}{\partial V}\right)_{T,N}=\frac{Nk_B T}{V-bn}-\frac{aN^2}{V^2}$$

This is the famous van der Waals equation of state for a real gas arrived at by Famous Dutch Physicist Johannes van der Waal who believed in the molecular theory of matter and imagined a gas to be made of molecules of finite size and that there is an attraction between these molecules. The biggest success of the van der Waal's equation of state was that it not only explained the deviation from the ideal gas law but also provided a reasonably correct description of the system near the gas liquid critical point and played a crucial role in understanding liquefaction of gases and phase transitions.

Isotherms

To appreciate the full consequences of van der Waal's equation of state it will be interesting to look at pressure versus volume plots for different fixed temperatures called isotherms. Conventionally, these isotherms are plotted, between $P=\frac{P}{P_c}$ and $\tilde{o}=\frac{V}{V_C}$ for different T′s measured in the units of T_C i.e. corresponding to $t=\frac{T}{T_c}$. Here $P_{c,}V_c$ and T_c are respectively the critical pressure, critical volume and critical temperature which we will look at after analyzing general features of the isotherms plotted in figure.

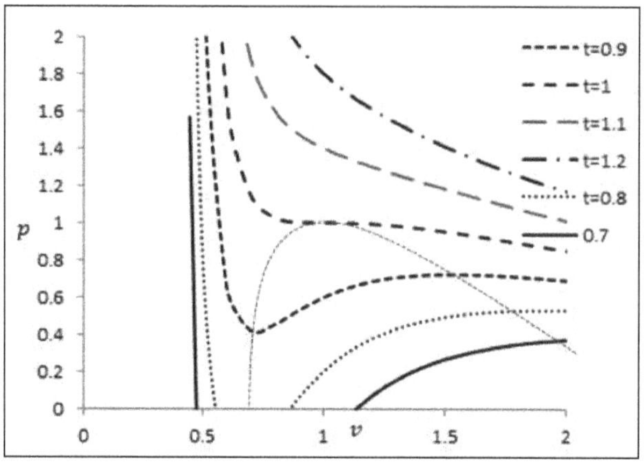

Isotherms of van der Waal's equation of state, red curve corresponds to the region where all isotherms have wrong sign (positive) of the slopes.

It is obvious from figure that above t = 1 curve i.e. for $T > T_c$, curves have the shape of ideal gas isotherms but below t = 1 i.e. for $T > T_c$, slopes of the curves for certain values of the volume and pressure become negative, which is an unphysical situation. Also stability condition demands that slope of the curves should always be negative, which is not true for some range of values of volume shown by drawing a curve joining the points where slope vanishes.

Condition for the vanishing of the slope is given by,

$$\left(\frac{\partial P}{\partial V}\right)_T = -\frac{Nk_B T}{(V-bn)^2} + \frac{2an^2}{V^3} = 0$$

For a given value of V, this condition is satisfied for a particular temperature called critical temperature (T_c) given by,

$$T_c = T = \frac{2aN(V-bN)^2}{Nk_B V^3}$$

Substituting this value of T in ($A = -K_B T \operatorname{In} Q_N(V,T) = -Nk_B T\left[\operatorname{In}\left(\frac{V-bn}{N\lambda}\right)+1\right] - \frac{aN^2}{V}$), we get

$$P = \frac{2abN^3}{V^3} + \frac{2N^2}{V^2}$$

A curve joining the points of zero slope on the isotherms is shown in figure 8, the top of this curve touches the isotherm corresponding to T_c called the critical temperature, signifying the fact that isotherms above this point all have negative slopes. Below T_c slope of a typical isotherm crossing this curve becomes zero then becomes positive and continues to be positive till it crosses the right boundary of the curve. On the left and right boundaries of the curve the second order derivative $\left(\frac{\partial^2 P}{\partial V^2}\right)_T$ is, therefore, respectively > 0 and < 0. As the temperature rises, the two boundaries merge till the curvature becomes zero at $T = T_c$. This implies that at $T = T_c$, both $\left(\frac{\partial P}{\partial V}\right)_T$ and $\left(\frac{\partial^2 P}{\partial V^2}\right)_T$ vanish corresponding to pressure and volume at this points representing critical volume V_c and critical pressure P_c. Equation $T_c = T = \frac{2aN(V-bN)^2}{Nk_B V^3}$ and the condition $\left(\frac{\partial^2 P}{\partial V^2}\right)_T = 0$ leading to:

$$\left(\frac{\partial^2 P}{\partial V^2}\right)_T = \frac{2Nk_BT}{(V-bN)^3} - \frac{6aN^2}{V^4} = 0$$

Which can be used together to calculate V_c. Once T_c and V_c are known, P_c can be calculated by putting these values in equation $\left(\frac{\partial P}{\partial V}\right)_T = -\frac{Nk_BT}{(V-bn)^2} + \frac{2an^2}{V^3} = 0$

Following this procedure, one gets,

$$V_c = 3Nb, T_c = \frac{8a}{27bk_B}, P_c = \frac{2}{27b^2}$$

Expressing P, Vand T in the units of P_c, V_c and T_c with $p = \frac{P}{P_C}, \upsilon = \frac{V}{V_c}$ and $t = \frac{T}{T_c}$ as dimensionless variables, the van der Waal's equation of state becomes,

$$p = \frac{8t}{3\upsilon - 1} - \frac{3}{\upsilon^2}$$

Equation in this form is a universal equation in the sense that it does not depend on the system parameters a and b explicitly, implying that units can be chosen for each system in such a way that form of the relation remains unchanged. This is law of corresponding states in real gases and is a result of similarity of interaction between the molecules of the gas.

Maxwell Construction

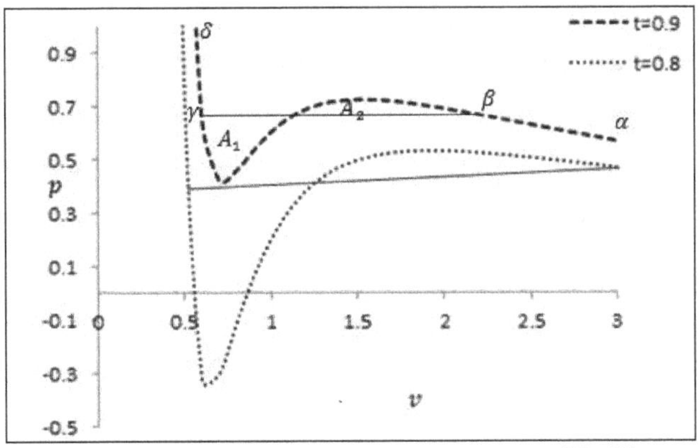

Maxwell construction on the isotherms for $T < T_c$.

The two problems discussed in the case of van der Waal's isotherms, namely:

- For < T_c, slopes of the curves for certain values of the volume and pressure become negative, which is an unphysical situation.
- Stability condition also demands that slope of the curves should always be negative.

Need to be sorted out. Solution to these problems was provided by Maxwell construction. To clarify the situation let us look at an isotherm of a typical gaseous system above T_c and below T_c as shown in figure.

Above T_c: for an isotherm of a gaseous system above T_c, let us decrease the volume. We notice that as volume decrease pressure decreases and we move smoothly from right hand side of the curve to the left hand side of the curve.

Below T_c: Below T_c, it will be nice to explain using a water as an example. Water has critical temperature of 647 K, let us pick this water system at temperature below T_c but slightly above its boiling point say 373 K. Also say the pressure is atmospheric pressure, temperature of the vapour is higher than the boiling point of water at this pressure.Now as we decrease the volume, pressure increases, simultaneously leading to increase in boiling point of water and hence condensation point is also higher. So as pressure increase water vapours start condensing. Since water occupies a much smaller volume on condensation for the same mass of water vapour, so when we try to decrease the volume further some vapour will condense further keeping the pressure constant. This implies that isotherms must have a horizontal region in the middle as shown in figure. For a given temperature at this pressure, liquid and water vapour co-exist. When whole of the water vapours get condensed, to decrease the volume further one needs to apply a very large pressure with a consequent steep rise in the isotherm. But what is the value of this constant pressure in the co-existence region described above. Maxwell provided an answer to this in so called Maxwell's construction which effectively means drawing a horizontal line below T_c cutting the isotherms at three points enclosing two bounded areas, one below the horizontal line on the left and other above the horizontal line towards the right and coexistence occurs at that pressure when the areas of these bounded regions become equal, resulting in a modified isotherm with a horizontal part. Hence there is now no part of the isotherm with positive slope and hence satisfies the stability criterion.

Let us look at above figure carefully, part of the isotherm δ to γ corresponds to liquid phase, part of the isotherm β to α corresponds to gas phase, part from γ to β corresponds to coexistence phase of liquid and gas states. Lowest volume intersection of the horizontal line with the isotherm gives volume of the liquid phase and the highest volume intersection gives the volume of the gas phase.

To justify the Maxwell's equal area construction idea further, let us recall the formula for Gibbs free energy at constant temperature:

$$dg = \upsilon dp$$

Integrating equation below along the isotherm, above figure, from α to β with integration over pressure the vertical axis, we have,

$$\int_{\alpha}^{\beta} dg = g_{\beta} - g_{\alpha} = \int_{\alpha}^{\beta} \upsilon dp$$

Or,

$$g_{\beta} - g_{\alpha} = \int_{\alpha}^{\beta} \upsilon dp$$

Similarly,

$$g_\gamma - g_\alpha = \int_\alpha^\gamma \upsilon dp$$

Thus,

$$g_\gamma - g_\beta = \int_\beta^\gamma \upsilon dp = A_2 - A_1$$

Thus if $A_2 > A_1$, then $g_\gamma > g_\beta$ and β will be the stable equilibrium, otherwise if $A_1 > A_2$ then $g_\gamma < g_\beta$, is the stable equilibrium. If $A_1 = A_2$ then the two phases can co-exist in equilibrium. This is also known as Maxwell's equal area law.

The most important success of the van der Waal's equation of state, which included attractive and repulsive part of the interaction between gas molecules lies in the prediction of gas-liquid transition.

Cluster Expansion

This is a method to calculate the higher order terms in the virial expansion. It introduces some general features of perturbation theory in many body systems, for example the use of diagrams to represent terms in the series, and the topology of diagrams to understand which ones have the correct extensive behavior that are therefore are the ones to be retained.

The configuration integral is written,

$$Z_N \int \ldots \int d^{3N} r e^{-\beta \sum_{i<j} u(|\vec{r}_i - \vec{r}_j|)} . = \int \ldots \int \Pi_{i<j} (1 + f_{ij}) d^{3N} r$$

with $f_{ij} = e^{-\beta u(r_{ij})} - 1$ with the idea that the effect of f_{ij} is small, either because the potential is small, or because a large potential only acts over a small volume, In this case we can expand the product in Z_{N},

$$Z_N = \int \ldots \int \left[1 + \sum_{i<j} f_{ij} + \sum_{\substack{i<j, k<l \\ i,j \neq k,l}} f_{ij} f_{kl} + \ldots \right] d^{3N} r,$$

where you can see what terms are in the multiple sums by thinking of all, unrepeated, interactions (e.g. 1–2 and 1–3 is included, but 1–2 and 1–2 is not). We then want to take logs to get thermodynamic quantities, e.g. in the canonical ensemble,

$$A = -k_B T \ln \left[\frac{1}{N!} \frac{1}{\lambda^{3N}} Z_N \right].$$

and in the microcanonical ensemble,

$$\frac{PV}{k_B T} = \ln Q = \ln \left[\sum_{N=0}^{\infty} \frac{1}{N!} \left(\frac{Z}{\lambda^3} \right)^N Z_N \right]$$

Note that the latter expression gives a convenient derivation of the virial expansion, since small density corresponds to small z. We will have to expand $\ln\left(1+a_1z+a2z^2\cdots\right)$ in small z and then relate ρ to z via,

$$\rho=\frac{z}{V}\frac{\partial \ln Q}{\partial z}.$$

The key idea is to represent the terms diagrammatically. This is generally a very useful approach in complicated perturbation expansions. Often simplifications of the results can be expressed in terms of the topology of the diagrams, such as the linked cluster expansion below. In the Mayer cluster expansion, each particle index is denoted by a dot or small circle, and each f_{ij} by a link between the dots i and j. Each diagram corresponds to a contribution to the expansion of ZN (with each term in the sums in Equation ($Z_N=\int\ldots\int\left[1+\sum_{i<j}f_{ij}+\sum_{\substack{i<j,k<l\\ i,j\neq k,l}}f_{ij}f_{kl}+\ldots\right]d^{3N}r,$) considered as a separate contribution), and there is an algebraic expression that can be associated with each diagram, as listed in the caption. The configuration integral Z_N is given by the sum of all distinct N-dot graphs. Note that in the counting, relabelled, similar-looking graphs such as in Figure are counted as distinct, although they will give the same expressions once the integrals are done. Since the integrals over the coordinates i, j for dots linked by f_{ij} gain contributions only from small r_{ij}, we think of the linked dots as particle clusters—there is a contribution to the configuration integral from all the various clusters that can form in the gas. Note the factorization property of unlinked clusters. Furthermore, each cluster has one free integration variable, giving a factor of V, with the remaining integrations over difference coordinates involving f giving factors of atomic volumes, so that the volume dependence of the whole graph is V^L with L the number of linked clusters.

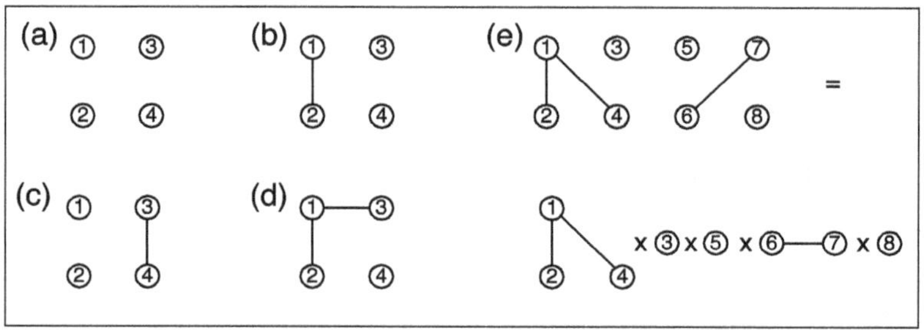

Some diagrams in the Mayer cluster expansion at fourth order representing: (a) $\int 1d^{12}r$,(b) $\int f_{12}\,d^{12}r$ (c) $\int f_{34}\,d^{12}r$; and (d) $\int f_{12}f_{13}\,d^{12}r$ (with $d^{12}r$ meaning $d^3r_1d^3r_2d^3r_3d^3r_4$. Note that (b) and (c) lead to the same values of the integral, and so are equivalent (but must be counted separately). Panel (e) shows an eighth order diagram demonstrating the factorization property: the value represented is the product of the terms $\int f_{12}f_{14}d^3r_1d^3r_2d^3r_4$, $\int d^3r_3$, $\int d^3r_5$, $\int f_{67}d^3r_6d^3r_7$, and $\int d^3r_8$.

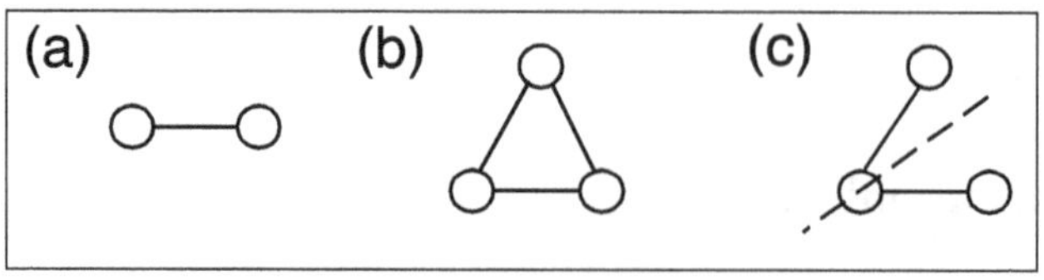

Linked clusters: (a) and (b) show the only irreducible, second and third order diagrams; (c) is a reducible third order diagram.

The diagrams are useful even for direct evaluation of low order expansions: as an example you will investigate the direct expansion up to third order in the homework. However the full power of the method comes because the complications introduced by the logarithm in forming the thermodynamic quantities can actually be eliminated by a general result that is most easily stated in terms of the diagrams, and in fact the expansions for the thermodynamic quantities are actually easier than for Z_N. This is the linked cluster expansion.

The linked cluster results take the form, for example for the grand canonical potential,

$$\ln \mathcal{Q} = \sum_{\substack{\text{linked}\\ \text{clusters}}} (\text{counting factor}) \times (\text{cluster integral})$$

where the summation is restricted to linked cluster, where a linked cluster is one with at least one f_{ij} line connecting each part of the diagram. The counting factor takes into account topologically similar diagrams that give the same integral, and numerical factors depending on the order of the term. Notice that we would guess that only linked clusters occur in the thermodynamic quantities, since other terms do not give the right extensive scaling with V. Finding out what the counting factor is, however, involves considerable careful work, which the references describe.

From this the virial expansion can be formed, and introduces a new topological property (irreducibility) of retained diagrams:

$$P = \rho k BT \left[1 - \sum_{l=1}^{\infty} \frac{1}{l+1} \beta_l \rho^l \right]$$

with

$$\beta_l = \frac{1}{l!} \times (\text{sum over all irreducible } l+1 \text{ clusters}),$$

where an irreducible cluster is one that cannot be cut into two pieces by cutting at a single dot. Reducible diagrams factor into simpler ones, since the integrals only depend on difference coordinates, e.g. for the diagram in Figure above(c).

$$\int\int f(r_{12}) f(r_{13}) dr_{12} dr_{13} = \int f(r_{12}) dr_{12} \int f(r_{13}) dr_{13}$$

(The particular form of Equation ($P = \rho k BT \left[1 - \sum_{l=1}^{\infty} \frac{1}{l+1} \beta_l \rho^l \right]$) comes from developing the expansion for ln γN with $\gamma N = Z_{N-1}/Z_N$.) The l = 1 term in the virial expansion in figure above(a) is what we have found before; the l = 2 term in equation $P = \rho k BT \left[1 - \sum_{l=1}^{\infty} \frac{1}{l+1} \beta_l \rho^l \right]$ depicted in Figure above(b) is

$$l = 2 : -\frac{2\rho^2}{3} \frac{1}{2!} \int \left(e^{\beta u(r_{12})} - 1\right)\left(e^{\beta u(r_{13})} - 1\right)\left(e^{\beta u(r_{23})} - 1\right) d^3 r_{12} d^3 r_{23}$$

The Liquid State

The classical liquid state is one of the most difficult states of matter to understand quantitatively, since there are no small parameters to use as a crutch (temperature, kinetic, and potential energies are all comparable). Quantum liquids are actually easier to understand, based on expansions in the number of thermal excitations at low temperatures (and the fact that the noninteracting gases already have interesting properties). A low order truncation of the virial expansion is not much use in understanding the strongly interacting classical liquid. There are two general classes of methods that are used to go beyond this. Neither of these methods are controlled approximations in general (nor are they particularly imaginative), but rather represent a last-ditch hope in the absence of really good methods. In some cases, e.g. with a long range Coulomb interaction, these methods may become exact in some limits.

If we focus on the deriving the equation of state, the calculation can be reduced to finding the radial distribution function $g_N(r)$. Remember that for the nearly ideal gas the radial distribution function is,

$$g\ (r)\ \ e^{-\beta u(r)}$$

We can phrase the calculation in terms of calculating W (r) defined by the correction to this,

$$g_N(r)=e^{-\beta u(r)}e^{W(r)}$$

or

$$\text{In}\, g_N(r)+\beta u(r)=W(r)$$

Usually, approximate expressions for W(r) are derived as functions or functionals of $g_N(r)$ leaving algebraic or integral equations to be solved for g_N.

Infinite Summation

If the low order Mayer (virial) expansion is of no use, we might try summing an infinite number of diagrams. The diagrammatic expansion for W is given by,

$$e^{W(r)}=e^{\beta u(r)}g_N(r)=\frac{V^2}{Z_N}\int\cdots\int\prod_{\substack{i<j\\ i,j\neq 1,2}}(1+f_{ij})d^3r_3\ldots d^3r_N$$

where the product on the right hand side does not include $(1 + f_{12})$ which gives the term $e^{\beta u(r)}$ on the right hand side. The diagrammatic expansion is similar to the Mayer expansion, except there are now two fixed points $\vec{r}_1,\vec{r}_2$ that we do not integrate over and provide "anchors" for the diagrams. It is not possible to sum all the diagrams, so various infinite classes are summed, chosen by some simplicity of topological form, or intuition of important physical content. The diagrams left out are in principle just as large as the ones retained, and so getting a good answer (for a particular result) is more an art than a science. Two examples are the "hypernetted chain" and the "Percus-Yevick" approximations.

Truncated Hierarchies

In this method a differential equation for $g_N(r)$ is derived. This equation depends on the next higher order density correlation functions. These in turn can be related to higher ones, and an exact hierarchy of equations can be derived. This is then truncated at some order, by making a simple factorization approximation of a high order correlation function, i.e. supposing that it factorizes into products of lower order ones. The hope is that high-order correlation functions do not have much physical content. Unfortunately, in most cases it seems that they do contain important physics, but this is now obscured in the complexity.

We define an effective "potential of mean force" φ_N (cf. the nearly ideal gas expression) in terms of the exact (unknown)g_N

$$\phi_N(r) = -k_B T \ln g_N(r)$$

and try to calculate the effective mean force,

$$-\frac{\partial \phi_N(\vec{r}_1, \vec{r}_2)}{\partial \vec{r}_1} = \frac{k_B T}{g_N}\frac{\partial g_N(\vec{r}_1, \vec{r}_2)}{\partial \vec{r}_1}.$$

(Although φ and g only depend on the difference $\vec{r}_1, \vec{r}_2$ the notation in terms of individual position is convenient here.) This can be interpreted as the force on the particle 1 at $\vec{r}_1$ holding particle 2 fixed at $\vec{r}_2$, but averaging over all positions of the other particles in the potentials of these two and their own interactions. Using,

$$g_N(\vec{r}_1, \vec{r}_2) = \frac{V_2}{N(N-1)} \frac{\int d^{3N}r \sum_{i \neq j} \delta(\vec{r} - \vec{r}_i)(\vec{r}' - \vec{r}_j) e^{-\beta U}}{\int d^{3N} r e^{-\beta U}}$$

$$-\frac{\partial \phi_N(\vec{r}_1, \vec{r}_2)}{\partial \vec{r}_1} = \frac{\int d^{3N} r_3 \ldots d^3 r_N \partial U / \partial \vec{r}_1 e^{-\beta U}}{\int d^3 r_3 \ldots d^{3N} r_N{}^{-\beta U}}.$$

The potential energy derivative $\partial U/\partial \vec{r}_1$ contains the term $\partial u(r_{12})/\partial \vec{r}_1$ involving the fixed particle and N – 2 terms of the form $\partial u(r_{1j})/\partial \vec{r}_1$ with $j \neq 1, 2$. Equation above can then be written as,

$$\frac{\partial \phi_N(r_{12})}{\partial \vec{r}_1} = \frac{\partial u(r_{12})}{\partial \vec{r}_1} + \rho \frac{\int d^3 r_3 g_N^{(3)}(\vec{r}_{12}, \vec{r}_{13}) \partial u(r_{13}) / \partial u(r_{13}) / \partial \vec{r}_1}{g_N(\vec{r}_{12})}$$

where $g_N^{(3)}$ N is the three-particle distribution function,

$$g_N^{(3)}(\vec{r}, \vec{r}') = \frac{V^2}{N(N-1)(N-2)} \left\langle \sum_{i \neq j \neq k} \delta(\vec{r} + \vec{r}_i - \vec{r}_j) \delta(\vec{r}' + \vec{r}_i - \vec{r}_k) \right\rangle$$

which can be used to define a third order effective potential $\phi_N^{(3)} = -k_B T \ln g_N^{(3)}$. This procedure can be repeated to define the hierarchy. We can truncate the hierarchy at second order by making the superposition or factorization approximation,

$$\phi_N^{(3)}(\vec{r}_{12}, \vec{r}_{13}) \simeq \phi_N(\vec{r}_{12}) + \phi_N(\vec{r}_{23}) + \phi_N(\vec{r}_{13})$$

$$g_N^{(3)}(\vec{r}_{12}, \vec{r}_{13}) \simeq \phi_N(\vec{r}_{12}) g_N(\vec{r}_{23}) g_N(\vec{r}_{13})$$

to give,

$$\frac{\partial\phi_N(r_{12})}{\partial\vec{r}_1}=\frac{\partial u(r_{12})}{\partial\vec{r}_1}+\rho\int d^3r_3\, g_N(r_{13})\frac{\partial u(r_{13})}{\partial\vec{r}_1}$$

This equation is usually expressed in the integrated form. First introduce the function,

$$E(r)=\int_{\infty}^{r}\frac{du(r')}{dr'}g_N(r')dr'.$$

Equation $\frac{\partial\phi_N(r_{12})}{\partial\vec{r}_1}=\frac{\partial u(r_{12})}{\partial\vec{r}_1}+\rho\int d^3r_3\, g_N(r_{13})\frac{\partial u(r_{13})}{\partial\vec{r}_1}$ reduces to,

$$\frac{\partial\phi_N(r_{12})}{\partial\vec{r}_1}=\frac{\partial u(r_{12})}{\partial\vec{r}_1}+\rho\int d^3r_3 g_N(r_{23})\frac{\partial E(r_{13})}{\partial\vec{r}}$$

The integral can then be expressed in the form,

$$\phi_N(r_{12})=u(r_{12})+\rho\int d^3r_3 g_N(r_{23})E(r_{13})+C.$$

The integration constant C is evaluated by taking particle 2 to infinity for fixed $\vec{r}_1,\vec{r}_3$($(\phi_N(r_{12})\to 0, u(r_{12})\to 0, g_N(r_{23})\to 1)$. The expression can then be written in the form known as the Yvon-Born-Green equation,

$$\text{In}\, g_N(r)+\frac{u(r)}{k_B T}=-\frac{\rho}{k_B T}\int E(|\vec{r},\vec{r}'|)[g_N(r')-1d^3r']$$

Notice this is an integral equation for gN in the form of Equation $\text{In}\, g_N(r)+\beta u(r)=W(r)$.

Quantum Cluster Expansion

This is an advanced topic to show how cluster expansions can be done for a dilute quantum gas. You can read it or skip it as you like. The more interesting aspects of quantum systems, such as Bose-Einstein condensation cannot be treated using this formalism, and the methods introduced later in the term are more useful in general.

The starting point is,

$$\mathcal{Q}=\sum_{N=0}^{\infty}Q_N(V,T)z^N=\sum_{N=0}^{\infty}\frac{Z_N(V,T)}{N!}\left(\frac{z}{\lambda^3}\right)^N$$

With $z=e^{\beta\mu}$ the fugacity, and where the second expression just defines the quantity Z_N (which in the classical case is the configuration integral). In the $V\to\infty$ we get,

$$\frac{P}{k_B T}=\frac{\text{In}\mathcal{Q}}{V}=\frac{1}{\lambda^3}\sum_{N=0}^{\infty}b_1 z^1$$

with

$$b_1 = \frac{Z_1}{V} = 1$$

$$b_2 = \frac{1}{2!\lambda^3 V}\left(Z_2 - Z_1^2\right)$$

$$\vdots$$

Note that Z_N scales as V^N, but the combination $Z_2 - Z_1^2$ appearing in b_2 must scale as V (cf. the linked cluster result), and similarly for higher order terms.

We can also calculate the density ρ,

$$\rho = \frac{z}{V}\frac{\partial \ln Q}{\partial z} = \frac{1}{\lambda^3}\sum_{l=1}^{\infty} l b_l z^l$$

which can be inverted to give a series for $z(\rho)$, and so with Equation($\frac{P}{k_B T} = \frac{\ln Q}{V} = \frac{1}{\lambda^3}\sum_{N=0}^{\infty} b_l z^l$) the virial expansion,

$$\frac{P}{\rho k_B T} = \sum_{l=1}^{\infty} a_l(T)\left(\rho\lambda^3\right)^{l-1}$$

With $a_1 = b_1 = 1$, $a_2 = -b_2, a_3 = 4b_2^2 - 2b_3 \ldots$.

The nth term in the virial expansion is given in terms of the partition functions for m particles with $m \leq n$. Classically, it is straightforward to write down expressions for Z_n and this is just reproducing the results of the Mayer expansion. Quantum mechanically, for Z_n we have to solve the n-body quantum mechanical problem, and this can only be done easily up to $l = 2$, i.e. the first nontrivial virial coefficient.

It is easiest to compare with the noninteracting system (that even the noninteracting quantum system has nonzero higher terms in the virial expansion),

$$b_2 - b_2^{(0)} = \frac{1}{2\lambda^3 V}\left(Z_2 - Z_2^{(0)}\right)$$

$$= \frac{\lambda^3}{V}\left(Q_2 - Q_2^{(0)}\right)$$

$$= \frac{\lambda^3}{V}\mathrm{Tr}\left(e^{-\beta H_2} - e^{-\beta H_2^{(0)}}\right)$$

With,

$$H_2 = -\frac{\hbar^2}{2m}\left(\nabla_1^2 + \nabla_2^2\right) + u(r_{12})$$

The eigenfunctions of the two body problem can be separated,

$$\Psi = \frac{1}{V^{1/2}} e^{i\vec{P}.\vec{R}/\hbar} \psi_n(\vec{r}_{12})$$

where ψ_n satisfies,

$$\left\{ -\frac{\hbar^2}{2\mu} + \nabla_{\vec{r}}^2 + u(r) \right\} \psi_n \vec{r} = \varepsilon_n \psi_n(\vec{r})$$

with $\mu = m/2$ the reduced mass, and then

$$b_2 - b_2^{(o)} = \frac{\lambda^3}{V} \sum_j e^{-\beta P_j^2/4m} \sum_n e^{-\beta\varepsilon_n} - e - \beta\varepsilon_n^{(o)}$$

$$= 8^{1/2} \sum_n \left(e - \beta\varepsilon_n - e - \beta\varepsilon_n^{(o)} \right)$$

where the sum over the center of mass momentum states is just $8^{1/2}V/\lambda^3$. The final sum over the energy eigenstates can be split into contributions from the bound states and from the continuum (scattering) states.

$$b_2 - b_2^{(o)} = 8^{1/2} \sum_B e^{-\beta\varepsilon B} + \frac{8^{1/2}}{\pi} \sum_l (2l+1) \int_o^\infty e^{-\beta\hbar^2 k^2/2m} \frac{\partial n_l(k)}{\partial k} dk$$

with n_l the phase shift. The l-sum is over l even for Bosons and l odd for Fermions.

Virial Coefficients

Temperature Dependence of Second Virial Coefficient: Universal Behaviour

Second virial coefficient given in equation ($B_3(T) = -\frac{1}{3} \int \left[f_{12} f_{13} f_{23} d^3 r_{12} d^3 r_{13} \right]$) is a function of temperature, T, alone. A typical form of the potential u(r) is given in figure below. For a typical r_o in the graph, if $r < r_o$, u(r) is large and positive leading to a positive contribution to $B_2(T)$. For values of r $< r_o$, u(r) is negative. Then for lower values of temperature, $\left(e^{-\frac{u(r)}{kBT}} - 1 \right)$ becomes large and positive, making $B_2(T)$ negative.

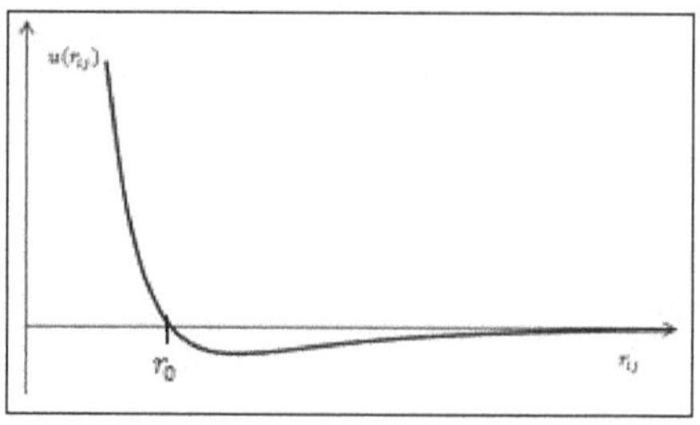

Typical pair potential.

If we take potential of the form $u(r) = -\in v(r)$, then $\frac{\in}{k_B T} \gg$ is low temperature regime and $\frac{\in}{k_B T} \ll$ 1 is the high temperature regime.

Also $B_2(T)$ can be cast in a reduced form, $B_2(T)$ reduced to $\frac{B_2(T)}{(2\pi r_o^3)/3}$ and $k_B T$ gets reduced to $k_B T/\in$. A graph of reduced virial coefficient and reduced temperature is shown in figure below. We observe that in reduced units virial coefficient of all noble gases fall on the same curve showing a universal behavior except He which is a light gas in which quantum effects become prominent.

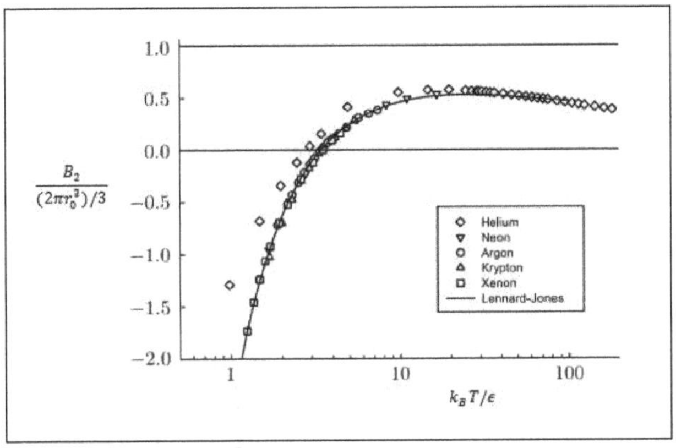

Figure above $\frac{B_2(T)}{(2\pi r_o^3)/3}$ versus $\frac{k_B T}{\in}$ ϵ for noble gases, except He which is a light quantum gas all other gases show a universal behaviour.

Virial Expansion of Equation of State for Different Model Potentials

In the following we now evaluate the virial coefficients for some well-known model potentials for a given system which is not very far from ideal gas behavior. Out of the three model potentials chosen here Lennard Jones potential is the one which reproduces universal behavior of real gases very nicely and is the most preferred model potential for carrying out computer simulations.

Hard Sphere Potential: Second and Third Order Virial Coefficient

The hard sphere potential is given by,

$$u(r) = \begin{cases} o \text{ for } r > r_o \\ \infty \text{ for } r \le r_o \end{cases}$$

Then the Mayer function becomes,

$$f(r) = \left(e^{-\frac{u(r)}{k_B T}} - 1 \right) = \begin{cases} o \ \text{ for } r > r_o \\ -1 \text{ for } r \le r_o \end{cases}$$

$$\text{using } \left(B_2(T) = -2\pi \int r^2 \left(e^{\frac{-u(r)}{k_B T}} - 1 \right) dr = -2\pi \int r^2 f(r) dr \right)$$

$$B_2(T) = -2\pi\int r^2\left(e^{-\frac{u(r)}{k_BT}} - 1\right)dr = 2\pi\int_0^{r_0} r^2 dr = \frac{2\pi r_0^3}{3}.$$

It is worth noting that second virial coefficient is independent of temperature as is expected.

The Third virial coefficient is given by,

$$B_3(T) = a_3(T)(\lambda^3)^2 = -\frac{1}{3}\int\left[f_{12}f_{13}f_{23}\, d^3r_{12}d^3r_{13}\right]$$

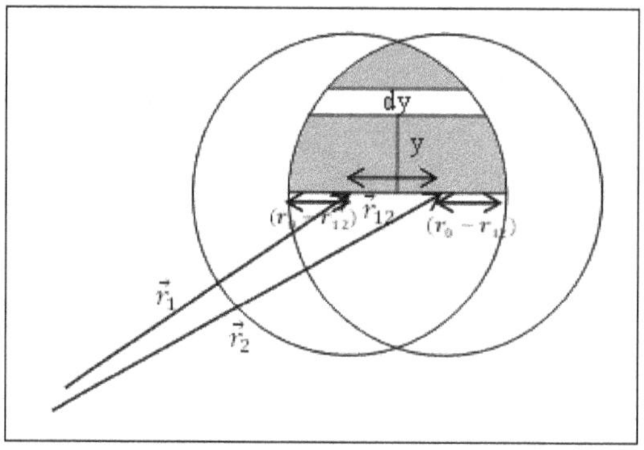

Region of integration for evaluating integral in $B_3(T)$.

We again recall that Mayer function f(r) = −1 for $r < r_0$ and zero for $r > r_0$. So to evaluate the integral in above equation we first fix the positions of the particles 1 and 2 described by the position vectors $\vec{r}_1$ and $\vec{r}_2$ in figure and allow particle 3 to take all possible positions so that integration over r_{13} may be performed. The integrand is -1, when $r_{13} < r_0$ and $r_{23} < r_0$ for all other values of r_{13} and r_{23} it is zero.

Therefore, ($B_3(T) = -\frac{1}{3}\int\left[f_{12}f_{13}f_{23}d^3r_{12}d^3r_{13}\right]$) can be written as,

$$B_3(T) = a_3(T)(\lambda^3)^2 = \frac{1}{3}\int_{r_{12}=0}^{r_0} d^3r_{12}\left(\int d^3r_{13}\right)_{common}$$

Where $\left(\int d^3r_{13}\right)$ *common* is the region common between the two spheres centered at $\vec{\ }$ and $\vec{r}_2$ respectively. The region over which this condition is satisfied is the volume enclosed by a complete rotation of the shaded region in the figure about the line joining the centers of the two spheres. To evaluate this let us cut a strip of thickness dy at y, which is of length $2\sqrt{r_0^2 - y^2} - r_{12}$. This strip sweeps out an an elemental area of $2\pi y\ dy$ and hence a volume of,

$$\left(\int d^3r_{13}\right)_{common} = 2\pi\int_0^{\sqrt{\left(r_0^2 - \left(\frac{r_{12}}{2}\right)^2\right)}}\left(2\sqrt{r_0^2 - y^2} - r_{12}\right)y\,dy$$

This can be solved easily as follows:

$$\left(\int d^3r_{13}\right)_{common} = 2\pi\int_0^{\sqrt{\left(r_0^2 - \left(\frac{r_{12}}{2}\right)^2\right)}}\left(2\sqrt{r_0^2 - y^2} - r_{12}\right)y\,dy$$

Putting $w=2\sqrt{r_o^2-y^2}-r_{12}, \frac{dw}{dy}=-\frac{4y}{\left(2\sqrt{r_o^2-y^2}-r_{12}\right)+r_{12}}=-\frac{4y}{w+r_{12}}$, that is $-(w+r_{12})\frac{dw}{4}=ydy$ limits change from $2r_o-r_{12}$ to o,

$$\left(\int d\ r_{13}\right)_{common}=\frac{2\pi}{4}\int^{2r_o\ r_{12}}(w+r_{12})wdw=\frac{\pi}{2}\left(\frac{w^3}{3}+r_{12}\frac{w^2}{2}\right)\Bigg|^{2r_o\ r_{12}}$$

$$=\frac{}{2}\left[\left(\frac{(2r_o-r_{12})^3}{3}+{}_{12}\frac{(2r_o-r_{12})^2}{2}\right)\right]$$

$$=\frac{}{2}(2r_o-r_{12})\left[\frac{2r_o\quad r_{12}}{3}+\frac{12}{2}\right]$$

$$=\frac{}{12}(2r_o-r_{12})\ [4r_o-2r_{12}+3r_{12}]$$

$$=\frac{16}{12}\left(r_o-\frac{r_{12}}{2}\right)\left[r_o+\frac{r_{12}}{4}\right]$$

$$=\frac{}{}\left(r_o-r_o\,r_{12}+\frac{r_{12}}{4}\right)\left({}_o+\frac{r_{12}}{4}\right)$$

$$=\frac{}{3}\left(r_o^3-r_o^2\,r_{12}+\frac{r_o\,r_{12}^2}{4}+\frac{r_o^2r_{12}}{4}-\frac{r_or_{12}^2}{4}+\frac{12}{16}\right)$$

$$=\frac{4}{3}\left(r_o^3-\frac{3}{4}r_o^2\,r_{12}+\frac{12}{16}\right)$$

i.e.

$$\left(\int d^3r_{13}\right)_{common}=\frac{4\pi}{3}\left(r_o^3-\frac{3}{4}r_o^2r_{12}+\frac{r_{12}^3}{16}\right)$$

So,

$$B_3(T)=a_3(T)(\lambda^3)^2=\frac{16\pi^2}{9}\int_{r_{12}=o}^{r_o}r_{12}^2\left(r_o^3-\frac{3}{4}r_o^2r_{12}+\frac{r_{12}^3}{16}\right)dr_{12}$$

$$=\frac{16\pi^2}{9}\left(r_o^3\frac{r_{12}^3}{3}-\frac{3}{4}r_o^2\frac{r_{12}^4}{4}+\frac{r_{12}^6}{96}\right)$$

$$=\frac{16\pi^2}{9}\left(\frac{r_o^6}{3}-\frac{3r_o^6}{16}+\frac{r_o^6}{96}\right)$$

$$=\frac{16\pi^2}{9}\left(\frac{32r_o^6-18r_o^6+r_o^6}{96}\right)$$

$$=\frac{16\pi^2}{9}15\frac{r_o^6}{96}=\frac{5\pi^2r_o^6}{18}$$

Here we can note an interesting relationship between third virial coefficient and second virial coefficient of a hard sphere gas,

$$B_3(T)=\frac{5\pi^2 r_0^6}{18}=\frac{5}{8}\left(\frac{2\pi r_0^3}{3}\right)^2=\frac{5}{8}(B_2(T))^2$$

Also third virial coefficient of a hard sphere gas is independent of temperature.

Square Well Potential: Second Order Virial Coefficient

The square well potential is given by,

$$u(r)=\begin{cases}\infty \text{ for } r<r_0\\ -\epsilon \text{ for } r_0<r<\lambda r_0;\ \text{potential well} \quad \text{Here } \lambda \text{ is the range of the}\\ 0 \text{ for } r>\lambda r_0\end{cases}$$

Then the Mayer function becomes,

$$f(r)=\left(e^{-\frac{u(r)}{k_B T}}-1\right)=\begin{cases}-1 \text{ for } r<r_0\\ \left(e^{\frac{\epsilon}{kT}}-1\right) \text{ for } r_0<r<\lambda r_0\\ 0 \text{ for } r>\lambda r_0\end{cases}$$

Using $\left(B_3(T)=-\frac{1}{3}\int\left[f_{12}f_{13}f_{23}d^3r_{12}d^3r_{13}\right]\right)$

$$B_2(T)=-2\pi\int r^2\left(e^{-\frac{u(r)}{k_B T}}-1\right)dr=2\pi\left[\int_0^{r_0} r^2 dr-\left(e^{\frac{\epsilon}{kT}}-1\right)\int_{r_0}^{\lambda r_0} r^2 dr+0\right]$$

$$=\frac{2\pi r_0^3}{3}\left[1-\left(e^{-\frac{\epsilon}{k_B T}}-1\right)\left(\lambda^3-1\right)\right]$$

In the case of square well potential we note that, second virial coefficient depends on temperature via the depth of square well potential. In the limit $T\to\infty$ $e^{\frac{\epsilon}{kT}}\to 1$ irrespective of the sign of ϵ and B_2 (T) is the same as that of a hard sphere potential. This is expected also, since particles of the gas have sufficiently high kinetic energy and are not affected by the potential well. At low temperature and for $\epsilon>0$, B_2(T)becomes large and negative. This means that particles have a tendency to bound together and hence exert less pressure than an ideal gas with same density and temperature.

Lennard Jones Potential: Second Virial Coefficient and Derivation of Van Der Waals Equation of State

Lennard Jones potential, a realistic potential for a real gas with an empirical foundation, is given by,

$$u(r)=4\epsilon\left[\left(\frac{\sigma}{r}\right)^{12}-\left(\frac{\sigma}{r}\right)^6\right]$$

It has the following characteristics:

- u(r) has a minima at $-\in$ at a distance $r_o = 2^{\frac{1}{6}}\sigma$.
- For $r < \sigma$ it rises to a large positive value.
- For $r \gg \sigma$ it approaches zero negatively.

Potential typically looks like figure. The portion to the left of the minimum is mainly repulsive which comes into play when particles come very close to each other, interaction to the right becomes attractive when particles start approaching each other from right. The second virial coefficient for the potential can then be obtained using ($B_3(T) = -\frac{1}{3}\int\left[f_{12}f_{13}f_{23}d^3r_{12}d^3r_{13}\right]$). Normally, it is convenient to work in reduced units by setting $x = \frac{r}{\sigma}$ and $T^* = \frac{k_BT}{\in}$ for reduced Lennard-Jones potential can be written as,

$$B_2^*(T) = -\frac{B_2(T)}{2\pi\sigma 3} = \int x^2\left(e - \frac{4(x-12-x-6)}{T^*} - 1\right)dx$$

However, qualitatively, the effective Lennard Jones potential can be written as,

$$u(r) = \begin{cases} \infty \text{ for } r < r_o \\ -u_o^{\left(\frac{r_o}{r}\right)^6} \text{ for } r \geq r_o \end{cases}$$

The Mayer function then becomes for this case,

$$f(r) = \left(e^{\frac{u(r)}{k_BT}} - 1\right) = \begin{cases} -1 \text{ for } r < r_o \\ e^{\frac{u_o}{kT}\left(\frac{r_o}{r}\right)^6} - 1) \text{ for } r \geq r_o \end{cases}$$

The second virial coefficient can then be written as,

$$B_2(T) = 2\pi\left[\int_0^{r_o} r^2\,dr + \int_{r_o}^{\infty}\left(e^{\frac{u_o}{kT}\left(\frac{r_o}{r}\right)^6} - 1)\right)r^2\,dr\right]$$

At higher temperatures in the limit $\frac{u_o}{k_BT} \ll$, the second integral can be simplified giving,

$$B_2(T) = 2\pi\left[\int_0^{r_o} r^2\,dr + \int_{r_o}^{\infty}\left(\frac{r_o}{r}\right)^6 r^2dr\right] = 2\pi\left[\frac{r_o^3}{3} - \frac{u_o r_o^3}{3kT}\right] = \frac{2\pi r_o^3}{3}\left[1 - \frac{u_o}{kT}\right]$$

Though this result is approximate, it reproduces van der Waals equation of state of a real gas. Recalling $\frac{N}{V} = \frac{1}{v}$ let us rewrite ($B_2(T) = 2\pi\left[\int_0^{r_o} r^2\,dr + \int_{r_o}^{\infty}\left(\frac{r_o}{r}\right)^6 r^2dr\right] = 2\pi\left[\frac{r_o^3}{3} - \frac{u_o r_o^3}{3kT}\right] = \frac{2\pi r_o^3}{3}\left[1 - \frac{u_o}{kT}\right]$)

$$\left(P+\frac{2\pi r_o^3}{3}u_o\left(\frac{1}{y^2}\right)\right)\cong\frac{k_BT}{v}\left[1+\frac{2\pi r_o^3}{3}\frac{1}{v}\right]$$

$$\left(P+\frac{2\pi r_o^3}{3}u_o\left(\frac{1}{y^2}\right)\right)\left(v\left(1+\frac{2\pi r_o^3}{3}\frac{1}{v}\right)^{-1}\right)\cong k_BT$$

$$\left(P+\frac{2\pi r_o^3}{3}u_o\left(\frac{1}{y^2}\right)\right)\left(v-\frac{2\pi r_o^3}{3}\right)\cong k_BT$$

If we choose $a=\frac{2\pi r_o^3}{3}u_o$ and $b=\frac{2\pi r_o^3}{3}$, we get

$$\left(P+\frac{a}{v^2}\right)(v-b)\cong Nk_BT$$

Here the parameter b is four times the actual molecular volume, which is volume of a sphere of diameter r_o.

References

- Classical-System-of-Interacting-Particles-I: inflibnet.ac.in, Retrieved 10, May 2020
- Dense-Gas-and-the-Liquid-State, Statistical-Mechanics: cmp.caltech.edu, Retrieved 18, February 2020

Permissions

All chapters in this book are published with permission under the Creative Commons Attribution Share Alike License or equivalent. Every chapter published in this book has been scrutinized by our experts. Their significance has been extensively debated. The topics covered herein carry significant information for a comprehensive understanding. They may even be implemented as practical applications or may be referred to as a beginning point for further studies.

We would like to thank the editorial team for lending their expertise to make the book truly unique. They have played a crucial role in the development of this book. Without their invaluable contributions this book wouldn't have been possible. They have made vital efforts to compile up to date information on the varied aspects of this subject to make this book a valuable addition to the collection of many professionals and students.

This book was conceptualized with the vision of imparting up-to-date and integrated information in this field. To ensure the same, a matchless editorial board was set up. Every individual on the board went through rigorous rounds of assessment to prove their worth. After which they invested a large part of their time researching and compiling the most relevant data for our readers.

The editorial board has been involved in producing this book since its inception. They have spent rigorous hours researching and exploring the diverse topics which have resulted in the successful publishing of this book. They have passed on their knowledge of decades through this book. To expedite this challenging task, the publisher supported the team at every step. A small team of assistant editors was also appointed to further simplify the editing procedure and attain best results for the readers.

Apart from the editorial board, the designing team has also invested a significant amount of their time in understanding the subject and creating the most relevant covers. They scrutinized every image to scout for the most suitable representation of the subject and create an appropriate cover for the book.

The publishing team has been an ardent support to the editorial, designing and production team. Their endless efforts to recruit the best for this project, has resulted in the accomplishment of this book. They are a veteran in the field of academics and their pool of knowledge is as vast as their experience in printing. Their expertise and guidance has proved useful at every step. Their uncompromising quality standards have made this book an exceptional effort. Their encouragement from time to time has been an inspiration for everyone.

The publisher and the editorial board hope that this book will prove to be a valuable piece of knowledge for students, practitioners and scholars across the globe.

Index

www.ingramcontent.com/pod-product-compliance
Lightning Source LLC
LaVergne TN
LVHW080918210926
848485LV00004B/672
9781639894963